A CENTURY OF ENTHUSIASM

MIDAS, NEVADA 1907 - 2007

Dana R. Bennett

Midas A CENTURY OF ENTHUSIASM Nevada 1907–2007

FRIENDS OF MIDAS
This publication was made possible through the generous support of Newmont Mining Corporation

NEWMONT™ The Gold Company

Dedication

This book is dedicated to the Midas School (November 1927 to November 2005) and all of the children and adults who learned there, worked there, played there. A true Nevada treasure, this beautiful building will continue to shine in countless happy memories.

This book was made possible by a generous contribution from Newmont Mining Corporation.

NEWMONT™
The Gold Company

All proceeds from the sale of this book go to Friends of Midas, a Nevada non-profit corporation with 501(c)(3) tax-exempt status. Friends of Midas is dedicated to the preservation of the history of Midas, Nevada.

Midas A Century of Enthusiasm Nevada
1907–2007

ISBN 0-9790127-0-8

First printing 2007

Additional copies may be ordered from Friends of Midas, HC66 - 75, Midas, Nevada 89414-9801
www.midasnevada.org

Printed in China

Page 1: Gus Bundy took this shot of Midas from the original road into town around 1947. (Photo courtesy of University of Nevada Reno Special Collections)

CONTENTS

Midas, Nevada, on a summer's day in the late 1970s (Photo courtesy of Roy Powers)

FOREWORD

Baby Midas Shaft Uncovers Body of High Grade Ore Filled With Free Gold!

A DISTRICT OF MOST WONDERFUL GOLD SHOWINGS

THEY ARE COMING FROM ALL OVER TO GOLD CIRCLE

IT IS FORWARD WITH ENTHUSIASM; WILL CELEBRATE FOURTH OF JULY

Radiating excitement, these headlines from *The Gold Circle News* are as current today as when they were first printed in the brand-new town of Midas 100 years ago. The Midas Mine is one of Nevada's major gold and silver producers in the 21st century, and people still travel long distances to visit the residential center of the Gold Circle Mining District, especially for the popular Fourth of July festivities.

This small town in the southwest corner of Elko County, far from any major Nevada city and remote by any standard, has generated enthusiasm and excitement since James McDuffy struck gold there in June 1907. A century later, the town that has never been empty is home — both literally and figuratively — to hundreds of people. Most may not be able to claim Midas as their primary residence (permanent residents now number less than two dozen), but they are quick to claim Midas as the home of their hearts. Tony Bennett (no relation to the author, by the way) may have left his heart in San Francisco, but many of us left ours in Midas!

I have been hanging around Midas since 1968 and curious about its stories from the moment I met F. Wayne "Willie" Wilcox. Willie showed up in Midas in 1978, 40 years after he graduated from eighth grade at the Midas School, on his

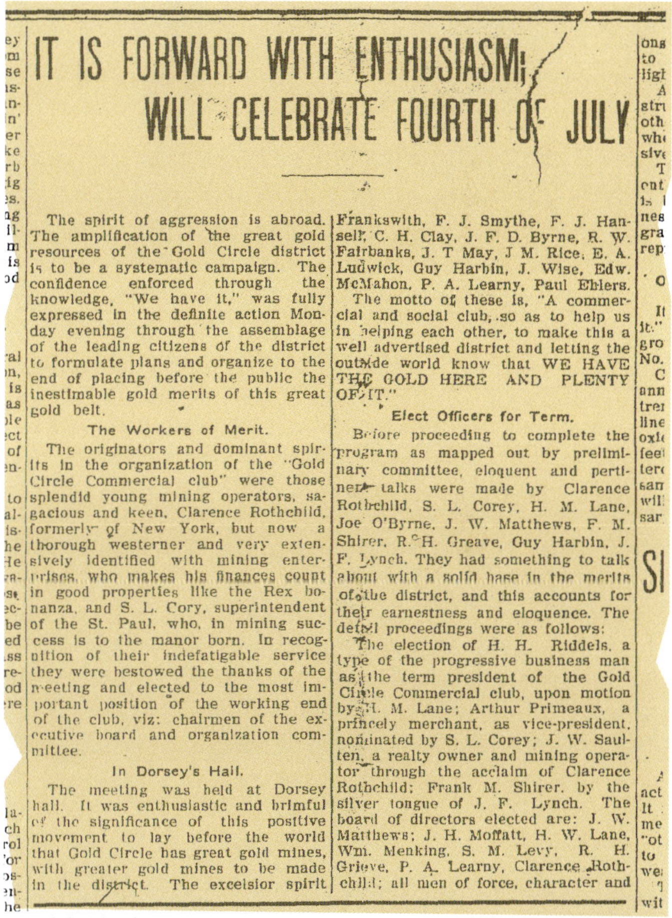

IT IS FORWARD WITH ENTHUSIASM; WILL CELEBRATE FOURTH OF JULY

The spirit of aggression is abroad. The amplification of the great gold resources of the Gold Circle district is to be a systematic campaign. The confidence enforced through the knowledge, "We have it," was fully expressed in the definite action Monday evening through the assemblage of the leading citizens of the district to formulate plans and organize to the end of placing before the public the inestimable gold merits of this great gold belt.

The Workers of Merit.

The originators and dominant spirits in the organization of the "Gold Circle Commercial club" were those splendid young mining operators, sagacious and keen, Clarence Rothchild, formerly of New York, but now a thorough westerner and very extensively identified with mining enterprises, who makes his finances count in good properties like the Rex bonanza, and S. L. Cory, superintendent of the St. Paul, who, in mining success is to the manor born. In recognition of their indefatigable service they were bestowed the thanks of the meeting and elected to the most important position of the working end of the club, viz: chairmen of the executive board and organization committee.

In Dorsey's Hall.

The meeting was held at Dorsey hall. It was enthusiastic and brimful of the significance of this positive movement to lay before the world that Gold Circle has great gold mines, with greater gold mines to be made in the district. The excelsior spirit

Frankswith, F. J. Smythe, F. J. Hansell, C. H. Clay, J. F. D. Byrne, R. W. Fairbanks, J. T May, J M. Rice; E. A. Ludwick, Guy Harbin, J. Wise, Edw. McMahon, P. A. Learny, Paul Ehlers.

The motto of these is, "A commercial and social club, so as to help us in helping each other, to make this a well advertised district and letting the outside world know that WE HAVE THE GOLD HERE AND PLENTY OF IT."

Elect Officers for Term.

Before proceeding to complete the program as mapped out by preliminary committee, eloquent and pertinent talks were made by Clarence Rothchild, S. L. Corey, H. M. Lane, Joe O'Byrne, J. W. Matthews, F. M. Shirer, R. H. Greave, Guy Harbin, J. F. Lynch. They had something to talk about with a solid base in the merits of the district, and this accounts for their earnestness and eloquence. The detail proceedings were as follows:

The election of H. H. Riddels, a type of the progressive business man as the term president of the Gold Circle Commercial club, upon motion by H. M. Lane; Arthur Primeaux, a princely merchant, as vice-president, nominated by S. L. Corey; J. W. Saulten, a realty owner and mining operator through the acclaim of Clarence Rothchild; Frank M. Shirer, by the silver tongue of J. F. Lynch. The board of directors elected are: J. W. Matthews; J. H. Moffatt, H. W. Lane, Wm. Menking, S. M. Levy, R. H. Grieve, P. A. Learny, Clarence Rothchild; all men of force, character and

Gold Circle News, June 20, 1908 (Courtesy of Nevada Newspapers Collection, University of Nevada Las Vegas Libraries)

first trip back to his treasured childhood town. I don't remember how we became acquainted, but he was soon regaling my sisters, Lori and Sheri, and me with wonderful tales of what it was like to be a child in 1930s Midas.

Here, in his own words, is one of his colorful vignettes:

> The gulch behind Primeaux's store and Eckman's Café was one of the "Outlaw Four" Gang's (Kenneth Fox, Frank Meyers, Glen Chipman, and Willie Wilcox) favorite gulch – summer and winter. In the gulch right behind Eckman's was an old one room corrugated steel shack. It had not been occupied for many years. So the four outlaws scrounged a door, windows, tables, chairs (powder boxes) and some old window blinds. This place was off limits to adults and girls. We would go by Hungry Joe's Saloon and see all those old guys inside playing cards and roulette. (No women.) So this was our model for our hideout. We gathered some empty whiskey bottles behind Joe's Saloon, cleaned them out and filled them with Kool Aid. The only card game we knew was "Go Fish." To complete this scenario we would cut up strips of newspaper and wrap sagebrush bark up and we had our booze and cigarettes. We would sit around our table, pour out a shot of Kool Aid, light up our roll your own cigarettes, take a shot of Kool Aid, and a puff on our sagebrush durhams and play Fish. Needless to say, our mouths tasted like burnt you know what.

Willie's stories came alive for me, and I could almost see and hear those 12-year-old boys building their clubhouse and coughing on their stogies. I am certain that Willie would happily accept the responsibility for revving up my interest in his beloved town.

For awhile, I haphazardly collected information and questioned all of the old-timers I could find. Edna Wilkerson Timmons, who had grown up in Midas and returned to retire, was a wealth of stories and pictures and was always happy to sit down with me in her Midas kitchen to chat about the old days. Lucky visitors to her kitchen often enjoyed three fingers of Wild Turkey with those tales! Her brother Byron Wilkerson, who was born in Midas, was a great storyteller in his own right, sprinkling his tales with more quips than, as he was fond of saying, "Gunny has sacks." Gordon Warren and his sister Desda Warren Wood, who were also born in Midas, patiently endured my interrogations as did Jack and Noreen Wilkerson Murdock, Bish and Loriene Macy Ferguson, Joe Keller, Evelyn Sabin, and others. Many of these folks have since passed away and are greatly missed.

At the prodding of several people, especially my uncle Dick Bennett, I started to write down what I had found, which encouraged me to become more systematic in my research. Yet I was surprised as anyone when I found I had enough material for a book. A decade ago, Great Basin Press published *Forward with Enthusiasm: Midas, Nevada, 1907-1995.* Contrary to all expectations, and again to my surprise, the 500 copies sold out quickly, leading to requests to re-print the book.

In the meantime, however, the mining industry returned to Midas. A mile from town is a world-class gold and silver mining and milling operation owned by Newmont Mining Corporation. A simple reprint of the book would not have included the newest chapter in the Midas mining story. Great Basin Press generously gave the publishing rights to Friends of Midas, the town's historic preservation group, and my father, Dan Bennett, suggested that I revise the book. Those of you who have been encouraged by him to support Friends

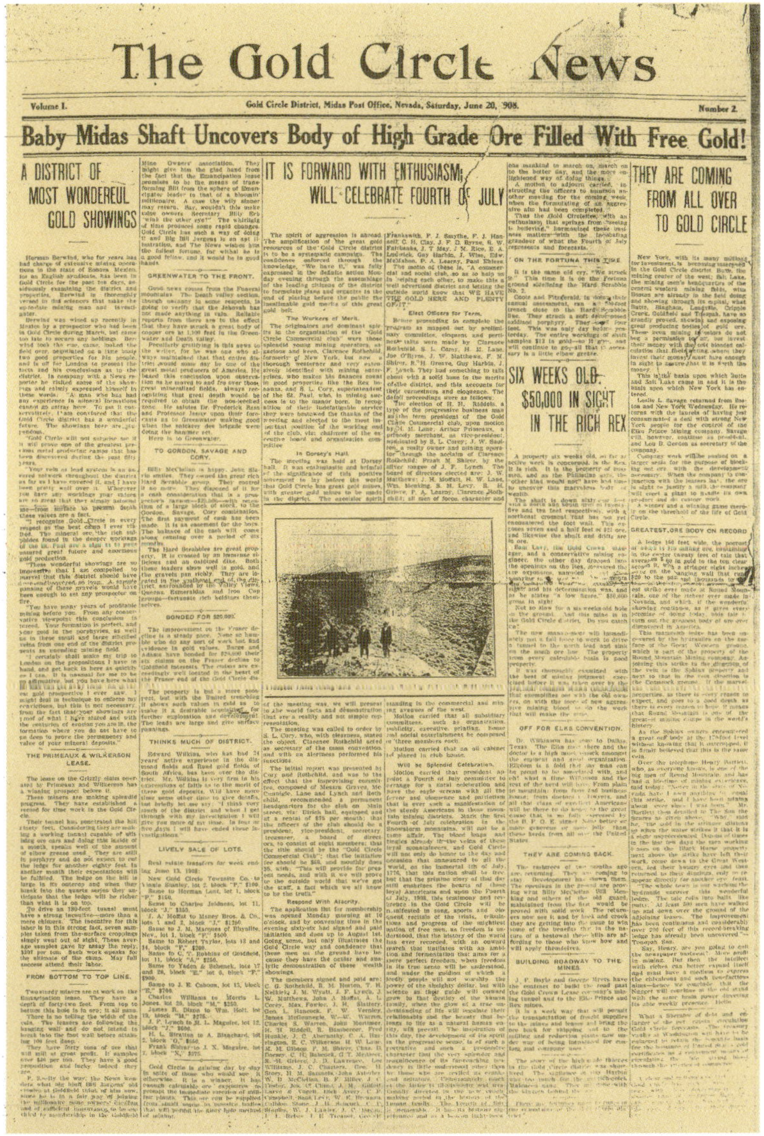

The Gold Circle News

Volume I. Gold Circle District, Midas Post Office, Nevada, Saturday, June 20, 1908. Number 2.

Baby Midas Shaft Uncovers Body of High Grade Ore Filled With Free Gold!

A DISTRICT OF MOST WONDERFUL GOLD SHOWINGS

IT IS FORWARD WITH ENTHUSIASM; WILL CELEBRATE FOURTH OF JULY

THEY ARE COMING FROM ALL OVER TO GOLD CIRCLE

SIX WEEKS OLD. $50,000 IN SIGHT IN THE RICH REX

GREENWATER TO THE FRONT.

TO GORDON, SAVAGE AND CORY

BONDED FOR $20,000.

THINKS MUCH OF DISTRICT.

THE PRIMEAUX & WILKERSON LEASE.

LIVELY SALE OF LOTS.

FROM BOTTOM TO TOP LINE.

The Workers of Merit.

In Dorsey's Hall.

Elect Officers for Term.

ON THE FORTUNA THIS TIME

Will be Splendid Celebration.

Respond With Alacrity.

OFF FOR ELKS CONVENTION.

THEY ARE COMING BACK.

BUILDING ROADWAY TO THE MINES.

GREATEST ORE BODY ON RECORD

Gold Circle News, June 20, 1908 (Courtesy of Nevada Newspapers Collection, University of Nevada Las Vegas Libraries)

A newly-arrived Midas enthusiast in 1907 or 1908 (Photo courtesy of Jerry Fenwick)

of Midas know how focused he can be on his task!

As the President of Friends of Midas, my father has done an amazing job of adding to the data that is available about Midas and the surrounding area. The archives belonging to Friends of Midas provide a treasure trove of stories and pictures. It quickly became apparent that much could be added to the original publication. With Midas completing its first century, the time was right for another book.

Consequently, the book that you now hold in your hands contains much of *Forward with Enthusiasm* and yet so much more – more information, more stories, and many more pictures. This book truly celebrates the first 100 years of Midas. It is a new book, not a second edition.

It is also organized differently, so a quick summary of the chronological history of the area is in order here.

The boom that resulted in Midas began in 1907. By 1908, fortune-seekers swarmed the strike area, which is at the base of the Owyhee Bluffs and adjacent to Squaw Valley approximately 50 miles northeast of Golconda and 40 miles southwest of Tuscarora. Two towns – Gold Circle and Summit – sprang up, and the Gold Circle Mining District was formed. But the deposits could not support the rush. In 1909, residents numbered less than 200, and Summit disappeared. Gold Circle was the only town, but the Federal Government changed the name to Midas. There is no indication that the town was ever called Rosebud.

After the initial rush, the mining district and the town limped along until large-capacity mills could be built. Late in 1915, the Elko Prince Mining Company began operation of its 50-ton cyanide mill, and production increased. For five years, the district was in its most productive period, second only to that begun in 1998, and the population again hovered around 200.

On April 15, 1922, a fire destroyed the Elko Prince mill, and the district headed into a depression. Two years later, Noble H. Getchell formed the Gold Circle Consolidated Mining

Company, which revitalized the district and the town for the next 20 years.

The date often given for the end of mining in Midas is 1942 when the Federal War Department closed many hard-rock mines across the country. In the 1950s, the University of Nevada ranked this district sixth in total production among 34 districts in Elko County. Although the official name of the district is Gold Circle, the ore contained a considerable amount of silver. Additionally, the district yielded copper, lead, and zinc.

After World War II, however, mining continued sporadically here, and the town endured. Midas is often found on lists of ghost towns, but it has never been completely abandoned. Since the 1960s, it has been a popular destination for hunters, fishers, and recreationists. Many people, including those who had lived here as children, have retired to Midas.

In 1998, underground mining began again, this time on a scale unimagined by earlier Midas miners. Over the district's first 90 years, 129,000 ounces of gold and 1.6 million ounces of silver were extracted. During the district's next ten years, more than 10 times as much gold and silver have been produced.

When the town was larger, Midas had saloons and restaurants, red-light districts, a jail, a town hall, and a first-class school house. Dances were popular; local elections mattered; and murders were committed on Main Street. But no church was ever built.

Today, Midas consists of a few older, wooden buildings and several new vacation homes along a tree-lined dirt road in a fairly narrow canyon. The town is a quiet place, populated by a handful of people year-round. The population increases exponentially on the weekends and holidays when the part-time residents arrive. The only business is George and Terri Watford's Midas Saloon and Dinner House. But the town has all of the comforts of modern life: electricity, an efficient water system, telephones, satellite dishes providing television reception, and access to the Internet.

Forward with Enthusiasm asserted that Midas was no longer a mining town. That is still true in the sense that the 226 men and women who work at the Midas Mine do not, for the most part, live in this town. When Midas Joint Venture began building the facility, company officials promised to minimize any adverse effects on the tiny community. Not only did they do that, but they, and the mine owners who followed them, helped wherever and whenever they could. These companies, past and present, include Barrick Gold Corporation, Newmont Mining Corporation, Normandy Mining Limited, Euro-Nevada Mining Corporation, and Franco-Nevada Mining Corporation; all made significant donations of time, labor, and funds to enhance local programs. They have been the key factors in the restorations of the Midas School and Midas Cemetery and the enhancement of the Midas Water Cooperative. In this sense, Midas is still very much a mining town.

A Century of Enthusiasm describes one hundred years of life and work in the mining town of Midas through pictures and stories. This book, and the one preceding it, however, would never have happened without the enthusiastic assistance of numerous Midas supporters and of organizations such as the Northeastern Nevada Museum, Elko County, and the North Central Nevada Historical Society.

First and foremost, though, a huge "thank you" must go to Newmont Mining Corporation for providing the funding for the design and printing of *A Century of Enthusiasm.* Newmont's generous assistance ensures that every dime you spent on this book goes directly to Friends of Midas. (Nope, no royalties for me, but you can buy me a rum-and-Diet-Coke® next time we're in Midas together!)

Newmont employees were also quick to answer my mining questions; my thanks to Mary Korpi, Edd Jackson, and Brent Kristof.

Nancy Peppin of Reno did a most amazing job designing this book. Not only did she create something of true beauty, but also she endured hundreds of e-mails from my dad and me with patience and good humor. No small feat, that!

A special acknowledgment goes to those who have, over the years, provided access to priceless community and family photographs, especially Len and Jeanne Berg, Anna Jossis Hinrichs (niece to Victor and Hilda Jacobson), Don Richardson, Phil Gemmill, Mary Lou Vest Smith, B.J. Matthews, and the Lyon sisters. Joyce Lyon Triplett and Minerva Lyon Dunn were both born in Midas to a family with long roots in this town. They are the granddaughters of Frank and Bell Macy and Isabel Lyon. Their father was Charles Lyon, and their mother was Minerva Macy, the first baby born in the new camp of Gold Circle in 1908.

Most of all, I must express my appreciation to my mom, Joan Bennett, who has been quietly indispensable to the success of Friends of Midas. Like many of the women who make modern homes in Midas, you won't see her name sprinkled across the official lists like the men's names are. But make no mistake: if it weren't for women like my mom, Paula Scheuller, Ida Unger, Bev White, and others I've known and admired my entire life ... if it weren't for these women who enthusiastically cook fabulous dishes for the potlucks, perform tedious administrative tasks, serve at community meals and clean up afterwards, and still find time to shoot pests, hammer nails, and wield a mean weed-eater, those men wouldn't get a darned thing done.

Finally, I extend my gratitude to everyone who told me, again and again, how much they enjoy *Forward with Enthusiasm.* I hope you like this book, too.

See you in Midas!

Dana R. Bennett
2006

Frank and Bell Macy (seen here in their 1906 wedding portrait) arrived in Gold Circle in October 1907. Their family grew with the town over the next several decades. (Photo courtesy of the Lyon Family, Friends of Midas Collection)

THE PEOPLE OF MIDAS

In 1907, during the first year of discoveries, many people rushed into the Gold Circle Mining District and into Midas. The proximity of Squaw Valley Ranch and the camp of Dutton to the find and the probable existence of other prospectors in the area ensured that James McDuffy and his sons were not alone for long in the new district. In fact, the Elko *Weekly Independent* reported on July 19, 1907, that approximately 60 people were at the discovery site. Within a few months, 250 people were in the district, according to the Elko *Free Press*, but more laborers were needed.

Two of those people were Frank and Edith Bell Macy, who arrived by horse and wagon on October 21, 1907, to join Bell's brother-in-law Charles Randall. For a few minutes every day of the next year, Bell took the time to jot down a few sentences in her "Day Book." On October 29, she counted 29 tents in camp and on December 1, 1907, noted that Mrs. Dorsey and she were "the only women folks in Gold Circle."

The number of people in the district continued to increase through the winter and spring of 1908. The Battle Mountain paper noted that the population doubled during February to 400, and more were arriving daily.

On May 12, 1908, the Elko *Free Press* indicated that 1,500 people could be found throughout the district. But, the paper noted, the influx of would-be millionaires had tapered off due to more excitement elsewhere in the West. The population had peaked. Fifteen months later, 200 men were working in the area. The following May, the 1910 census taker, Benjamin Trembath, counted 163 souls in town. Ten years later, the federal enumerator found 212 Midas residents.

Bell Macy kept a daily journal throughout her first year in Midas. (Photo courtesy of the Lyon Family, Friends of Midas Collection)

This bustling view of downtown Midas in February 1908 was transferred onto a postcard. (Photo courtesy of Jack and Noreen Wilkerson Murdock, Friends of Midas Collection)

The first big drop in population began after the Elko Prince Mill burned in 1922 and steadily decreased until only 25 people remained in town. But the arrival of Noble Hamilton Getchell in the Gold Circle Mining District in 1924 would ultimately result in a revitalization of the district.

Getchell's activities sparked a new interest in the district. With the announcement of the company's creation in 1926, boom times came again to Midas, and people began streaming back into town. The Battle Mountain *Scout* reported in October that "12 cars of freight consisting of machinery, lumber and supplies have been hauled into Midas during the last few weeks and 14 more cars are en route." By January 1927, the paper noted that the town's population had increased 14 times to 350 people. Those folks did not stay long, however.

In 1930, the census taker, Jessie Diamond, counted 130 people in Midas. After 1930, available census information does not list the population of each of the towns in Elko County. Consequently, the best information about the number of people over the age of 21 years who were in town during the later years comes from the election results. In 1936, 89 people cast their ballots in Midas for the general election; in 1940, 66. By the end of the decade, school records noted the enrollment of 20 children, indicating that the annual population of Midas was probably around 100.

The biggest decline in population began in 1942 when World War II, a mill fire, and decreasing mine production combined to drive away residents. In March 1942, the "Midas News Notes" reported that numerous people were moving out of the district; only eight children were attending the town's school. No one was moving into the area. The town's businesses began to terminate their operations, and most shut their doors forever. On September 16, 1942, the post office was closed permanently. Only 17 people voted in the November general election. By the end of 1942, the school was closed because it could not enroll at least three students.

Noble Hamilton Getchell
(Courtesy of the Special Collections Department, Getchell Library, University of Nevada, Reno)

Hungry Joe Horvath and Chris Christensen were long-time Midas residents. Note the campaign poster for Pat McCarran on the wall behind Christensen. (Photo courtesy of Jack and Noreen Wilkerson Murdock, Friends of Midas Collection)

Despite the twin declines of mining production and population, Midas did not die. Some people never left; eventually, new residents arrived. The school reopened in 1943 and mining activities persisted, but Midas never regained the population it once had. It was also never empty. The winter months of the 1960s would often find only one resident (Andy Jones) in town.

In 1952, a roaming ghost town chronicler counted nine residents, although Elko County recorded 21 registered voters in the town during August of that year. A reporter for the *Nevada State Journal*, recounting the story of Midas on March 30, 1958, noted that mining activities increased in the fall of 1957. Approximately 40 people moved into town, and mail delivery resumed, although the post office was not reopened. As with earlier booms, however, most of these people did not stay long: in 1958, 16 registered voters listed Midas as their residence. In 1960, the county recognized eight voters in the town; in 1970, 11. Over the next 40 years, the number of the full-time residents has remained relatively stable around 20, although the weekend population often exceeds 100 people.

Currently, Midas is mostly a retirement community. That trend began in 1972, when Edna Wilkerson Timmons, who had spent her childhood in Midas, and her husband Tim chose to retire in Midas. Others followed; some had also spent younger years in the town; some had discovered Midas as a recreational outlet. Although Newmont Mining Corporation operates a mine within a mile of Midas, few of those employees live in Midas. The company provides bus transportation between the mine and Winnemucca where most choose to live because of the availability of schools, stores, and other infrastructure that Midas no longer has.

Who Were These People?

The three census reports currently available (1910, 1920, and 1930) provide details about the people who lived in Midas. Without exception, Midas residents claimed their race as "white," and the vast majority of them spoke English as their primary language.

The 1910 census counted 98 men, 38 women, and 27 children. Men outnumbered women over two to one, and most (67 percent) were not married. Additionally, five married men were in Midas without their wives. Conversely, most women were married (74 percent), and no married woman was in town without her husband. Only 10 single women were in Midas; two of them were divorced, which was quite scandalous in 1910. The 163 residents occupied 91 households in Midas. The majority of household heads owned their homes free and clear. Only 35 percent of the citizens rented their homes.

But not all were U.S. citizens. Ten percent of the population (21) came from 10 different countries. Only one immigrant (from Ireland) had been naturalized by 1910. Other countries represented were England, Germany, Austria Hungary, Canada, Scotland, Finland, Sweden, and Spain.

Ten years later, not much had changed in Midas when Celestine Bache and Chester French arrived to conduct the 1920 census. Of the 190 people in Midas that January, 97 were men; 41, women; and 52, children. The ratio of men to women was still over two to one. Fifty-nine percent of the men were not married compared to 13 percent (a total of six) of the women. None of the women was divorced, but four married women were in Midas without their husbands. At least three married men were also without their spouses. Almost one-quarter of the men (24) lived together in a boardinghouse, one of the 71 households in the town. Forty-four of the household heads owned their homes free and clear, and twenty-six rented houses. Only one had a mortgage on a home.

The percentage of the population not born in the United

Cousins Willie Wilcox (left) and Glen Chipman (right) enjoy a Midas summer day with their friend Bonnie Richardson in 1938. (Photo courtesy of F. Wayne Wilcox, Friends of Midas Collection)

Gilbert Gemmill is ready for some Wild West action in 1959. (Photo courtesy of Phil Gemmill, Friends of Midas Collection)

In 1959, Bill Gemmill, his wife Evelyn (standing in yard), and their children lived where the John Elges home now stands. (Photo courtesy of Phil Gemmill, Friends of Midas Collection)

States had increased substantially. In 1920, over a third of Midas residents were foreign-born. Twenty-one residents listed their "mother tongue" as Swedish. Most of these people had immigrated from Finland, including Martin and Annie Brown. The other countries represented included Germany, New Zealand, England, Ireland, Spain, and Italy. In addition, the town's cobbler, Samuel Edelman, came from Russia and spoke Yiddish. Few of these immigrants had become U.S. citizens by 1920.

The 1930 census chronicled 70 men, 32 women, and 28 children. Over half (41) of the men were not married. This census detail included ranches; 30 men were employed in some facet of ranch work. By contrast, 28 were employed in mining work. Three men, such as Louis Reymond, listed their occupations as prospectors. Of the 64 households, 11 consisted of a male head, wife, and children. One household was composed of a widowed father and his four children; three others were mothers with their children. It was not unusual for men to live alone, but it was for women. In 1930 Midas, only five women lived alone. Two of them were the town's teachers; one, the postmistress. The only two divorced women—Marian B. Adams and Lois Griffith, each 42 years old—also lived alone. No occupation for either was recorded.

The percentage of homeowners decreased in 1930. Over half of the households rented; only 29 owned their own homes. Twenty-six Midas residents were born outside of the United States and represented 14 different countries. Seven were born in Spain (most were sheep herders and ranch hands), and five were born in Finland (the Brown and Berg families). The other countries represented were Italy, the Irish Free State, Switzerland, Portugal, Czechoslovakia, Belgium, England, Scotland, Austria, Germany, France, and Denmark.

A handful of people appear on all three census details: Katherine Boyle, Martin Brown, Annie Brown (on the 1910 census as Annie Sanders), Frank and Bell Macy, William

Muffley, Michael Paisley, Arthur Primeaux, William and Mary Rae, Frank and Katherine Raup, Herbert Thompson, Charles Warren, and Ruth Warren (on the 1910 census as Birda Holbert). Paul Ehlers, the "Lucky Dutchman" who found the Elko Prince Mine, was in Midas for the first two counts, but missed being included in the 1930 census by dying, at the age of 68 years, two months before the enumeration began.

Many of these people lived the rest of their lives in Midas. William Rae succumbed to a heart attack in his Midas home on April 6, 1934. Rae began his mining career in Cripple Creek, Colorado, and arrived in Midas soon after the first discovery in 1907. He leased several properties in the Gold Circle Mining District until his death.

Herbert Thompson lived in Midas until one week before his death on September 15, 1938. He had arrived in Midas in 1907, coming from Tuscarora and Cripple Creek, Colorado, before that. He was 73 years old.

Frank Macy died while still holding the office of Justice of the Peace in 1943. A year later, Martin Brown passed away in Midas where he operated a garage and rented cabins to hunters and other visitors. Annie Brown was too ill, confined to the Winnemucca hospital, to attend his funeral.

According to the obituary for Frank Raup, he and his wife had homesteaded a ranch about six miles from Midas in 1904. In addition, his obituary noted, "for some years he was in charge of the ball mill and solutions at the Midas mill when mines in that area were in operation." Frank died in 1950 in Elko where he was taken after being stricken at their ranch; Katherine died at the ranch a year later. Both were in their 80s.

There were other Midas residents who arrived in the early days of the first rush and stayed in the area for decades, but did not appear on all three census details for whatever reason. Three such residents arrived in 1907. Albert Geabbeart lived in Midas until his death on July 12, 1939, at the age of 84 years. Miriam Stenson Purdy arrived to run a boarding house and was still living there in 1952 when she was 73 years old. Chris Christensen called Midas home until he was admitted into the Elko hospital where he died on December 5, 1956, at the age of 72 years.

Martin and Annie Sanders Brown were Midas fixtures from the town's very beginning. (Photo courtesy of Edna Wilkerson Timmons, Friends of Midas Collection)

The Rex Mine operation, one of the longest-running mines in the Midas area, in 1911
(Photo courtesy of the Nevada Historical Society)

GOLD CIRCLE MINING DISTRICT

For much of the first 40 years of the 20th century, the Gold Circle Mining District, anchored by the town of Midas, was one of Nevada's top gold producers. After 1942, for over a half-century, no ore of any notice was taken out of any mine in the district, despite the belief of many prospectors and geologists that the area could still be productive, That belief was finally realized in 1999. In each of the following years, the Midas Mine (formerly known as the Ken Snyder Mine, formerly known as the Rex-Grande) produced more than all of its Gold Circle predecessors produced, in total, in 90 years.

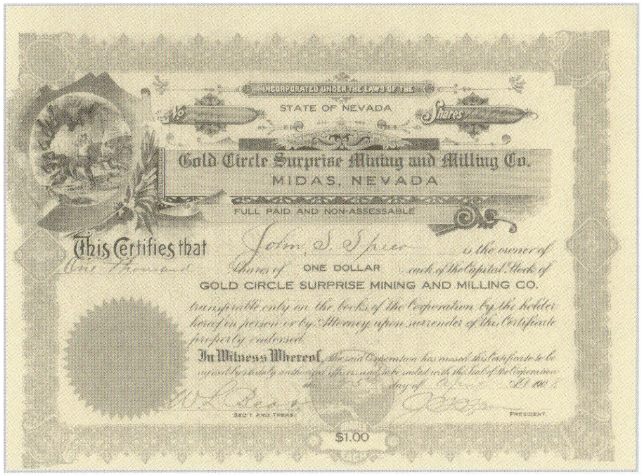

1907-1912: Finding Gold

The Gold Circle Mining District began in the spring of 1907 when James McDuffy and his two young sons were prospecting near Summit Creek in Elko County just east of the Humboldt County line. The area was not unknown; over the years, pannings in that location had yielded rich and promising gold flakes. Prospectors had been seeking the origin of this gold for some time without success. Then, during that trip, McDuffy made a major discovery. He found the source.

Although the newspapers did not report the find until that summer, word spread quickly among other prospectors and ranch hands already in the area. By June 11, 1907, several claims had been staked all around McDuffy's find, and a new mining district was formed.

According to newspaper articles, McDuffy called the district "Gold Circle" because "a perfect circle of malipi" surrounded the mineral belt, which he asserted was three miles long and

The Elko Prince was the largest mine in the area for many years. This shot of the mill and outbuildings was probably taken around 1920. (Photo courtesy of the Nevada Historical Society)

one mile wide. However, several claim notices and some newspaper reports called the district "Golden Circle," possibly indicating a slight disagreement over the name of the district. By the end of the year, the use of "Golden Circle" faded, and the name "Gold Circle" was permanently attached to the district.

On June 27, 1907, Battle Mountain's paper, the *Central Nevadan*, provided the first known report of the find. "Great Rush Now On In Squaw Valley Over Big Gold Strike," proclaimed a small article on the third page. "Claims Being Staked Off For Miles," the headline continued. Over 20 claims were staked and officially located during that month; by the end of the summer that number had tripled. The distinction of locating the first claim goes to the district's founder, McDuffy, and his partners, P.A. McKenzie and H.L. Bartlett. They located the Red Bank Lode No. 1 and the Black Jack Mine on June 12, 1907.

Paul Ehlers is mentioned by later publications as the founder of the district because he discovered the Elko Prince, which eventually became the biggest mine in the area, but it was not located until June 16, 1907. Ehlers, "The Lucky Dutchman," lived the rest of his life in Midas, passing away in 1930. His wife, Marie, and daughter, Edith, remained in the town for some years after his death.

Many of the earliest claims did not develop into significant mines, although they had some of the more colorful names: Buster Brown, Iron Dyke, Mohegan, Lone Rock, Daisy Lode, and Orphan Boy. Some of the claims that did become major producers were staked during that first summer, such as the Elko Prince; the Gold Crown, located on July 22, by William C. Menken and C.W. Schein; and the Sleeping Beauty, located on August 4, by George B. Barker and William Mears.

By September 1907, the Gold Circle Mining District consisted of 12 square miles of claims. *The Mineral Resources of the United States* stated that the district had "high values in gold. The ore is free milling and water is abundant . . . ledges [are] reported to be rich . . . placers are expected to be productive." The report, though generally optimistic, warned of the scarce wood and cold winters in the district. Predominant in the district was lode gold, but the deposits of silver, which were classified as the secondary metal in the area, soon rivaled gold production.

In October, Richard "Dick" Bamberger, who had re-located McDuffy's original claim, made the richest discovery up to that point. On October 20, 1907, the Tuscarora correspondent reported to the *Nevada State Journal* that:

> Many valuable nuggets and specimens of ribbon gold have been secured from the 2-inch high grade seam. Load after load of people have gone from Tuscarora and returned with pockets filled with valuable specimens and the supply seems almost inexhaustible. The resident of northern Elko county that does not carry a specimen of Gold Circle ore is considered an out-of-date and odd character.

Enthusiasm ran high among the fortune seekers in Midas that first year. There were at least two and usually more applications for every block of ground leased. At many of the claims, shafts were being sunk and tunnels dug. Before the end of 1908, one shaft was down 200 feet and several others were at the 100 foot level.

A year after the initial discovery was announced, the Elko *Daily Free Press* observed that the first shipment of ore had just left Gold Circle for Salt Lake City. Mined at the Gold Crown, this ore was sent to the

The Elko Prince Mine around 1920 (Photo courtesy of Friends of Midas)

refinery with much optimism and hope that it would be followed quickly by many additional shipments.

During the first few years, the mines in the Gold Circle Mining District all suffered from the same curse: the high cost of shipping ore out for processing. Until local mills could be built, it was not worthwhile to mine much ore. Work began quickly on local amalgamation mills, yet many mine owners were unable to pay their workers the going wage of $4 a day. The Elko Prince Mining Company warned its employees that work would be halted unless they agreed to receive company stock, then valued at 10 cents a share, as half of their wages. So, amid much scoffing by skeptics who saw the Elko Prince employees as suckers, the miners accepted these terms. The laughter subsided, though, when these miners eventually saw a return of at least 250 percent; some received as much as 500 percent on their stock. By 1916, the Elko *Daily Free Press* recorded the price of Elko Prince Mining Company stock at 65 to 75 cents a share.

In 1911, the *Mineral Resources* noted that two amalgamation mills were present, but inactive, and a 40-ton cyanide mill was processing ore. In 1914, most ore was treated at the Rex Reduction Works and came from the Rex, Colorado, Elko Prince, Hardscrapple, Midas, and Judge Mines. In 1915, the report explained that "motor trucks" were being used to haul ore in Nevada, but the use of engine-driven vehicles did not alleviate the costly problem of transporting ore. Unless the roads were in good shape (rare throughout the rest of the state and certainly not true of Midas roads at this time), the expense of using a truck was basically the same as using a wagon pulled by horses.

Despite transportation difficulties, the mines kept producing. Low-grade ore was processed at the local mills, such as Charles Geyman's amalgamation mill, and small quantities of high-grade ore were shipped out of the district. The Rex (location of a cyanide mill), the Elko Prince, and the Esmeralda Mines were the major producers from 1909 through 1914. In 1913, one of the biggest mines in the area was the Esmeralda, which employed 40 men. Additional mines, ranging in number from seven to 14, also reported operations during that period. The most successful year during this period was 1912 when production totaled 7,284 tons of ore yielding $59,784 from seven operators. Thirty-six pounds of copper were also produced.

Although the initial boom had ended, the newspapers occasionally printed glowing stories about the Gold Circle Mining District. In 1912, the Elko *Daily Free Press* reprinted an article from the *Nevada State Journal* proclaiming that pending deals "will mean great low grade mining employing thousand[s] of men in mining and reduction of ores." On October 22, 1912, the Elko paper recounted the discovery of a ledge of pure gold at the Eastern Star Mine. The find was so exciting that the mine's employees immediately requested a raise from $4.50 to $7 a day. When the owners refused, the men went on strike. The paper confidently noted that "as soon as the men get over the excitement, work will be resumed."

The Elko Prince buildings in 1912. (Photo courtesy of Lela Halsell Hempton, Friends of Midas Collection)

This new engine was the pride of the Elko Prince Mill in 1912. (Photo courtesy of Lela Halsell Hempton, Friends of Midas Collection)

1915 - 1923: A Prosperous Time

In June 1915, the Elko newspaper reported that the Elko Prince Mining Company filed suit in district court against the Golconda Cattle Company in order to obtain two miner's inches of water from Summit Creek needed to operate a mill that it planned to build. The suit must have been successful because the paper noted, four months later, that the area's mines were eagerly anticipating the opening of the mill. The 21 mines in the Gold Circle Mining District were in various stages of development and doing well, but all expected to improve as soon as the mill was working. On November 13, 1915, the Elko Prince Mining Company began operation of its 50-ton cyanide mill, and mining yields increased exponentially. The district entered a prosperous period that lasted almost a decade.

Production jumped by over 20,000 tons of ore from 1914 to 1916, and the yield increased by over $300,000. Over half of all the ore produced during the district's first 90 years of existence was mined between 1916 and 1921. During this five-year period, the mines produced over 104,000 tons of ore, yielding better than $1.7 million, making these years the most profitable and valuable in the district's history, until 1999. The majority of the ore came from the Elko Prince Mine and was treated at the Elko Prince Mill. Each year, that mine was the district's top producer. In 1917, *Mineral Resources of the United States* reported that the mine was also Elko County's most profitable producer.

At the same time the lawsuit was filed to obtain the necessary water, the Elko Prince Mining Company contracted with the Dorr Company to build the mill that would have an accelerating effect on the district's production. Operating through the Elko Prince Leasing Company, the Dorr Company retained control of the mill through December 1917, although the terms were completed in June of that year. Treating 50 tons of ore a day, the mill's lifespan was

This 1920 map illustrates the complexity of the mining claims filed in the Gold Circle Mining District. (Map courtesy of the Nevada Historical Society)

In 1928, Mr. and Mrs. Leslie Noble wrote from Midas: "For au [gold], we have quite a record." (Friends of Midas Collection)

promised to be over three years.

At the beginning of the operation, electricity was not available at either the mill or the mine, so a 10-horsepower oil engine was installed with a storage capacity of 22,000 gallons of oil. Water came from a well below Midas and had to be piped over a distance of 11,000 feet with a lift of 700 feet. However, electricity must have become available. In July 1916, the Midas correspondent to the Elko newspaper reported that the engine had been replaced by an electric motor, noting that "it is working fine and the noise of the exhaust of the gas engine is eliminated."

The Elko Prince Mill passed the three-year mark in November 1918. Despite some work delays in April and June of 1919 due to generator trouble, it appeared that high levels of production would continue indefinitely.

In 1920, the *Nevada State Journal* printed an article that provided an overview of the Gold Circle Mining District. It opened with the following:

> Perhaps the most surprising thing that is noted on one's first visit to Gold Circle is that a silver-gold camp could have so many mining properties of proved worth, or of more than ordinary promise, and yet be so little known, not only to the general public but to those who follow mining as a business. Without attempting to be poetical one feels that it is a neglected Cinderella in the sisterhood of precious-metal camps. This is all the more remarkable since the district is easily accessible ... The climate in summer is delightful and the winters are by no means severe as a rule.

The long, enthusiastic article went into great detail about the mining activities in the district, which was described as "no less than 29 mining claims on which ore ranging in value from milling grade to what the enthusiast refers to as 'jewelry' or 'picture' rock, has been found." Several sources indicate that Dick Bamberger gave to friends and neighbors chunks of gold found on his original site, and this article explained that the quality of Gold Circle ore was such that they actually did become jewelry, such as stick-pins and cuff links.

Then, early in the morning of April 15, 1922, a spark from a miner's lamp ignited the gasoline tank at the mill. The tank exploded, causing a tremendous fire. The entire mill burned to the ground.

Shortly thereafter, the Elko Prince Mining Company permanently ceased operations and began liquidation. The company determined that the ore remaining in the mine was not valuable enough to justify rebuilding the plant.

The opening of the mill had caused an explosion of production in the district. The explosion of the mill stopped the district in its tracks. In 1921, 12,733 tons of ore were produced, yielding $204,447; in 1923, 165 tons barely yielded $2,000. The district headed into a depression.

Might this have affected the fate of 42-year-old W. F. Boyken who had lived in Midas for five years? "One of the best known civil and mining engineers in the state," according to the *Reno Evening Gazette*, was found dead in his Midas cabin from an "overdose of sleeping powders" on February 21, 1923. The paper noted that "Boyken was heavily interested in the Big Chief Mining Company." He was buried in San Francisco.

The headframe at the Jackson Mine (Photo courtesy of the Northeastern Nevada Museum)

A closer view of the Jackson Mine (Photo courtesy of Friends of Midas)

1924-1935: The Heydays

The mines around Midas slowed production significantly. In November 1923, Dick Noble and his son Leslie sold the Colorado Grande to John H. Miller, the Republican state Senator from Mineral County. The *Reno Evening Gazette* explained that "the Nobles had taken out quite a mill run which they were going to put through the Elko Prince mill when that structure was destroyed by fire."

But it was the arrival of Noble Hamilton Getchell in the Gold Circle Mining District in 1924 that revitalized the district. Getchell, the Republican state Senator from Lander County

In 1931, the office of the Gold Circle Consolidated Mining Company was located at the intersection of the school road and the main road (where the Powell residence is now). (Photo courtesy of the Lyon Family, Friends of Midas Collection)

This view of the Grant Mine and Gold and Silver Circle Mill was taken around 1935. (Photo courtesy of the Lyon Family, Friends of Midas Collection)

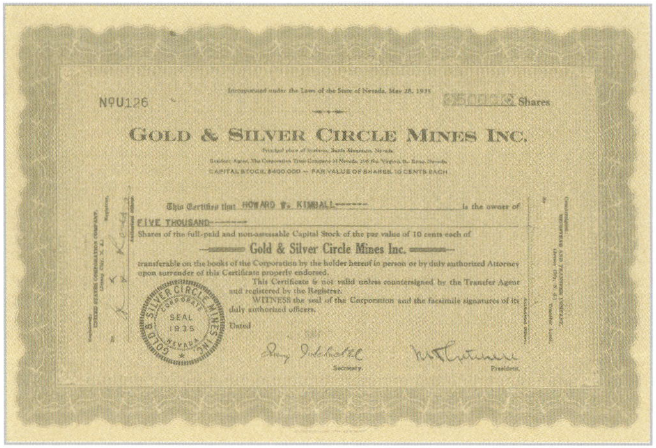

Incorporated under the Laws of the State of Nevada, May 28, 1935

Nº U126

Shares

GOLD & SILVER CIRCLE MINES INC.

Principal place of business, Battle Mountain, Nevada

CAPITAL STOCK, $400,000 — PAR VALUE OF SHARES, 10 CENTS EACH

This Certifies that HOWARD W. KIMBALL------ is the owner of

FIVE THOUSAND------

Shares of the full-paid and non-assessable Capital Stock of the par value of 10 cents each of

Gold & Silver Circle Mines Inc.

transferable on the books of the Corporation by the holder hereof in person or by duly authorized Attorney upon surrender of this Certificate properly endorsed.

This Certificate is not valid unless countersigned by the Transfer Agent and registered by the Registrar.

WITNESS the seal of the Corporation and the facsimile signatures of its duly authorized officers.

GOLD & SILVER CIRCLE MINES INC. CORPORATE SEAL 1935 NEVADA

Dated

Secretary. President.

(1923-1941), was a successful mining man whose primary operation in the early 1920s was the Betty O'Neal silver mine near Battle Mountain. On May 20, 1924, Getchell acquired the Elko Prince and Rex Mines after George Wingfield decided not to buy into the district. Apparently, Wingfield was not interested in repeating the investments he had made in the district 16 years earlier. According to the Battle Mountain *Scout*, Wingfield and the Goldfield Consolidated Company chose not to purchase the Elko Prince in 1924 without stating an official reason. The rumor was that they feared that the value of ore still in the mine decreased significantly near the bottom of the mine.

Also in 1924, Getchell and State Mine Inspector (and perennially unpopular candidate among Midas voters) Andy J. Stinson joined forces to re-open the Eastern Star Mine. According to Getchell's papers housed at the University of Nevada, Reno, Getchell utilized funds from Betty O'Neal

The Gold and Silver Circle Mill as it appeared around 1937. (Photo courtesy of Edna Wilkerson Timmons, Friends of Midas Collection)

stockholders to acquire other properties in the Gold Circle Mining District. These mines included the Missing Link, the Jackson Run, the Gold Slippers, the New Year, the General Grant, the Silver King, the St. Paul, and the Squaw Valley. On May 6, 1926, the Midas mines were legally joined with the Betty O'Neal to form the Gold Circle Consolidated Mining Company.

Although Getchell was optimistic about his newly acquired mines, he had some obstacles to overcome before they became productive. Two of these hurdles involved the tremendous amount of water in the Elko Prince Mine and the lack of a mill. By the time the mine was de-watered and a new, modern 75-ton cyanide mill built, Getchell and the Gold Circle Consolidated stockholders were $500,000 in debt. But Getchell's enthusiasm proved well-founded: within two years, the debt was retired, and the company began paying dividends. The Gold Circle Consolidated quickly became a major producer in Nevada. In the year before the company was created, the entire district produced just 61 tons of ore. Two years later, in 1927, the Gold Circle Consolidated was the ninth largest gold producer in the entire state. Each year after that to 1930, the company was ranked by the U.S. Department of Interior among the state's top eight or nine gold producers that, together, provided over 75 percent of the total amount of gold produced in Nevada.

The Gold Circle Mining District stumbled slightly going into the 1930s. In 1929, the Gold Circle Consolidated mill was shut down, pending additional mine development. Production fell from almost 14,000 tons of ore in 1929 to 552 tons in 1930. The Battle Mountain *Scout* (also purchased by Getchell in 1926) noted in July of 1929 that people were leaving the district to look for employment because some of the mines had temporarily closed.

But the downturn did not last long. Aided by President Franklin D. Roosevelt raising gold prices, the Gold Circle Mining District began again to produce significant amounts of gold and silver. In May 1930, the Battle Mountain *Scout* announced that Getchell had resumed operations in the mines of the Gold Circle Consolidated. The Elko Prince was being de-watered and its shaft deepened. Milling was expected to resume by the end of the year. The paper provided the following explanation for the brief shutdown:

> The property was closed down last year because ore in sight was practically exhausted and treasury funds were low. This was caused by the fact that much of the money derived from the mill was expended on the purchase price of the various mines now owned by the company.

By 1933, the Gold Circle Consolidated was, again, one of Nevada's major operations and one of the top six silver producers in the state. One of the company's holdings, the Elko Prince, was the top gold and silver mine in Elko County in 1935.

Other mines, such as the Eastern Standard (where Patrick Divine was the Superintendent) and the Miners Gold, also contributed to the success of the district, making it the biggest producer in Elko County in 1933 and 1934. On July 7, 1933, the Battle Mountain *Scout* reported that the Miners Gold had hit a significant vein and was piling up ore. The article indicated that high grade ore would be sent to Salt Lake City smelters and low grade to the local custom mill. The newspaper enthusiastically exclaimed that "this is the greatest strike in the district since the palmy days of the Elko Prince." The favorable report continued:

> The foregoing marks the opening of another mine in Nevada that has possibilities of being as exciting as the days of Goldfield. Midas is a busy town and there is not a house in town to be had. Capital is hunting properties and visitors are numerous.

Also by 1933, the Miners Gold Mine, owned by Charles S. Warren, was one of the top 14 gold producers in the state. In October of that year, the Midas columnist for the Battle Mountain paper reported that "Midas has had more mining activity in the last 10 days than has been here for years" and attributed the increase to production at the Miners Gold. At some point in 1933, Warren sold the mine to the Miners Gold Company whose officers included Edwin L. Veeter, President; J.A. Foley, Vice-President, Treasurer, and Manager; Earl Pemboke, Superintendent; J.A. Barclay, Secretary; and J.W. Ottenstein, Assistant Secretary.

A year later, the same individuals established the Midas King Gold Mining Company to operate the Bamberger Mine, which had, according to the Battle Mountain newspaper, "the record of producing the richest ore ever discovered in the Gold Circle District."

The *Reno Evening Gazette* trumpeted the 1934 strike made by Al Geabbeart and Tom Van Orman. "Old-Timers Strike Pay Dirt Near Midas," the paper proclaimed, giving a rather poetic view of the prospectors:

> They are two old sons of the desert, who have met with many rebuffs in their mining and prospecting labors in recent years, but who have always clung to the belief that they would strike it rich tomorrow, at least the next day.

According to the article, they were transporting their ore, which was primarily gravel, two miles away to Squaw Creek to run it through a Long Tom, a longer, bigger type of sluice box used in placer mining. The article concluded:

> It will not be long now before cold weather will freeze the gravel, forcing a suspension for the winter, but before that time, Van Orman and Gabbard [sic] expect to have enough in the kitty to keep them in comfort until the bitter blasts subside and the summer sun shines once more on their diggings.

Clearly, the reporter did not know that Geabbeart, 79 years old, had struck gold at least once before in the district. The *Nevada State Journal* announced, on November 11, 1907, a rich strike on claims owned by Dr. Edward Dunscomb and Geabbeart. However, the ground on which those claims were located did not become a mine. Instead, it was the location of the Dunscomb Townsite, which was never built. Both Dunscomb and Geabbeart lived out their days in Midas.

Despite positive news reports from the district, the Gold Circle Consolidated had begun to lose money, partly because of the failure of Nevada banks in 1932. According to Getchell's correspondence, the company was soon $82,000 in debt—mainly from keeping employees on the payroll despite lowered production—and Getchell had to find funding. Late in 1934, the company was reorganized into the Gold and Silver Circle Mining Company. The new organization was "taking over all the assets and liabilities of the Gold Circle Consolidated Mines, Betty O'Neal Mines, and Cash Mine," according to a letter from Getchell on October 25, 1934. A public offering of stock was made, and Getchell regained his enthusiasm for Midas. It was not to last long.

The mill at the Esmeralda Mine (Photo courtesy of Robert Pilant, Friends of Midas Collection)

A month later, Getchell wrote to a friend in San Francisco that the lack of water was impeding the use of the mill, "something that has never occurred before in the history of this camp." Storms had brought some, though not enough, moisture. The Elko Prince Mine had plenty of water but a 500-foot pipe was required to get the water to the mill. The company had neither the money nor the pipe to transport the water.

The lack of funds ended Getchell's involvement in Midas. In June 1935, the Battle Mountain paper's headline proclaimed: "Midas Rejoices As Payday Completed By Lander Senator." Getchell had just distributed paychecks from the Gold and Silver Circle Mining Company to pay off some of the debts of the Gold Circle Consolidated. According to Getchell's papers, he sold the majority of his stock to George Ormsby to pay employees who had been working for several months without receiving their wages. With the transfer of power to Ormsby, Getchell became dissatisfied with mining in Midas and soon turned his attention elsewhere. By July 18, 1935, he and his wife had moved permanently from Midas to Betty O'Neal.

According to the Battle Mountain *Scout*, the 1937 stockholders' report for the Gold and Silver Circle Mining Company explained Getchell's divestiture:

> Early last year, his Boston associates of many years having despaired of remaining in the mining business in the face of what they considered tyrannical SEC restrictions, N.H. Getchell, with a side line as a state senator but heading the roster of Nevada mine developers, disengaged himself from further involvements, became a free agent and opened thereafter the Getchell mine, 43 miles east of Winnemucca in Humboldt county.

1936 - 1942: Mining Tapers Off

Getchell was one of the most aggressive promoters that Midas had ever known. Certainly, his ownership of the Battle Mountain *Scout* provided the district with more press coverage than it had received since 1908. His affiliation with financial powers in San Francisco and New York City brought investment money into the mines and into the town. However, his departure from the area did not stop mining production or end the town's existence. In fact, the district enjoyed four more years of mining success.

In 1936, mining production almost tripled the output of 1935, and it increased yearly through 1939. During those four years, 36 percent more ore was mined than during the previous eight years. However, the total value was not as high: almost 90,000 tons of ore yielded over $573,000 from 1935 through 1939 compared to over $826,000 extracted from 66,500 tons during the preceding eight years. In March 1936, the "Midas News Notes" announced that "the mill has started full force and everyone is happy."

But not everyone *was* happy. On April 23, 1936, the Battle Mountain *Scout* reported that the miners employed by the Gold and Silver Circle Company threatened to strike if their wages were not increased by 50 cents to $5 per day. Ultimately, they accepted the company's offer of a daily increase of 25 cents and did not strike. The article noted that the leaders of the effort (four or five unnamed individuals) had departed the area, leaving the company short-handed.

A year later, Esmeralda Gold Mines, Ltd., was formed in Toronto, Canada, to acquire 16 claims and the Buena Vista mill on 240 acres from the original holders, brothers John W. and Robert T. Pilant. The Midas staff for the Canadian company included Mill Superintendent Victor Jacobson, Mine Superintendent H.L. Heath, and Chief Electrician A.C. Tracy. Mrs. Mae McKenna operated the boarding house. Development was progressing satisfactorily, including the installation of a 50-ton cyanide mill, until Canada entered World War II in 1939. At that time, export of Canadian capital was prohibited, resulting in the company going into American bankruptcy. Don Burgner, a Reno contractor and mine operator, purchased the company's assets from the federal bankruptcy court in January 1946. Canadian investment in Midas would return 50 years later.

In 1938, the *Scout* reported that the 75-ton custom mill was still operating, but was processing more dump ore than newly-mined ore. Dump ore from the Elko Prince Mine was treated daily. Occasionally, mined ore from the Colorado Grande and Rico Mines was also worked. Despite the shift to dump ore processing, enthusiasm for the area continued. In February 1938, the *Scout* described the negotiations underway to finance a 500-ton cyanide mill for the East Standard Mining Company. The mine was one mile south of Midas between the Elko Prince and Esmeralda and consisted of the Sleeping Beauty claims. Henry Wilcox had been hired as the superintendent of the East Standard Mine in 1936 and brought his wife, Lounella, and son, Wayne, to Midas from Utah. He held that position until he was laid off in December 1938, and the family left the area.

Earlier in 1938, Harry S. Joseph, president of the Salt Lake City-based East Standard Mining Company, explained that the company operated a 40-ton amalgamation mill, but the mine had a "tremendous body of low grade ore" that could not be

Re-construction began quickly after the 1941 fire in the Gold and Silver Circle's powerhouse. (Photo courtesy of Edna Timmons, Friends of Midas Collection)

treated profitably by amalgamation. Rich, high grade ore, which the mill was designed to treat, appeared to have been mined completely. The company planned to conduct further exploration and build additional mine buildings. Joseph indicated that the company's high production was expected to continue, but, in reality, the mill was never built.

In March, the *Scout* printed two stories concerning planned improvements to Midas mines. First, an article noted that a 35-ton cyanide mill was ready to be hauled to the Esmeralda mine as soon as the weather would permit. Crews for the mill and mine were already being hired. The second story, a front-page article about the stockholders' report, stated that "although economic difficulties have been experienced for the last several years at the Gold and Silver Circle Mines, Inc., property near Midas, a brighter outlook is seen for 1938." The report included positive forecasts for the Betty O'Neal and Midas properties, noting that "the company's highly efficient 75-ton cyanide mill" was busy treating ore from the Elko Prince Mine. In fact, the mill was running at capacity with Elko Prince dump ore. The company expected to operate the entire summer with ore from the Elko Prince and "other mines now being extensively developed near Midas."

The *Reno Evening Gazette* reported, in January of 1938, that the leasers were keeping the district in business, despite some recent setbacks. Identified as Joe Horvath, Ed Wilkerson, Jim Anthony, Chris Christianson, Charles Lyon, Alfred Scambi, Harry Lauritzen, and Jack Murdock, the leasers on the Colorado Grande had hit a pocket of gold ore and were now "making better than wages."

One year later, the *Scout* continued its positive reporting of the main industry of Midas, citing "mining activity on the increase" with two mills working full shifts and "numerous operators developing promising properties." The two operating mills were the Gold and Silver Circle cyanide mill, under the supervision of Claude W. Touzeau, general manager and president of the Gold and Silver Circle Mines, and the Esmeralda mill, which was "newly repaired" and "recently reconstructed with a cyanide unit." Six months later, the paper indicated that the Eastern Star, inoperative for some time, was starting up again. Other mining activities included the processing of 75 tons of ore daily by the Gold and Silver Circle plant, the preparation of ore for shipment by the Miners Gold, the operation of the Esmeralda mill, and the shipment of ore from the Queen to the East Standard mill.

In June of 1941, the *Scout*, still optimistic about mining in Midas, noted that the Eastern Star property rebounded and had been in "constant operation" for over a year; that summer, a night shift was added. In addition, both the Esmeralda and Gold and Silver Circle mills had reopened after repairs. It appeared that the strong production enjoyed since 1927 would continue indefinitely.

One of the first indications that problems existed had appeared on May 7, 1938, when the machinery of the Eastern Star Mine was sold at a sheriff's sale. Then, in 1940, the amount of production began a downhill slide. Tonnage decreased by over 20,000 tons from the previous year and dropped another 2,000 tons the next year.

Additional problems plagued the district. The *Scout* reported that fire ravaged the powerhouse of the Gold and Silver Circle Mining Company early in the morning of May 18, 1941, forcing the milling operations to

Chris Gemmill (top) and Dolores Gemmill (bottom) at the Esmeralda Mine in 1968. (Photo courtesy of Phil Gemmill, Friends of Midas Collection)

cease. Apparently, the fire started from a short circuit in the main switchboard of the powerhouse at 2 a.m. and quickly "jumped to the oil-splattered floor and swept the galvanized iron building." The company estimated the damages at less than $500 and began immediate repairs. By July, the plant reopened and began to treat "considerable ore on a 24-hour basis."

But the Gold and Silver Circle mill was not open long. On July 1, 1942, *The Humboldt Star* reported that a sheriff's sale was held at the company's office in Midas for all of the property, buildings, and milling machinery of the Gold and Silver Circle Mines, Inc., to pay off a mortgage of $110,000. Evidently, the company had been in bankruptcy since 1939. Its properties included the Elko Prince, Colorado Grande, Rico, Rex, Missing Link, Jackson, and Toothpick Mines.

Three separate bids totaling $3,750 were received, far below the amount needed. The article indicated that the Elko sheriff would present the bids to Elko District Court, which would probably reject them because they did not come within $105,000 of the needed funds. According to the paper, the mill, including all of the machinery and the building, received a bid of only $1,000. Bids for the remaining property included $450 for the powerhouse, $150 for the office building and equipment, and $110 for the Getchell house. The next day, the *Star* reported that the Battle Mountain State Bank (for which Getchell was a director) made a blanket offer of $10,000 for all of the properties, equipment, and buildings. This bid was first rejected, then accepted.

In 1942, the Federal War Production Board issued order L-208, effective on October 15, which closed gold mines throughout the country. The action was intended to conserve labor and materials needed for World War II and specifically affected only lode mines that produced more than 1,200 tons of ore in 1941. No mine in the Gold Circle Mining District produced that much ore in 1941; thus, although the order is often blamed for shutting down the district, it actually did not affect the area's mines.

Instead, it appears that the Gold Circle Mining District suffered a natural demise. High-grade ore, never the bulk of the production in the district, had become even more rare by 1942. Low-grade ore, which was more common in Midas mines, was also becoming scarce. Although the mines still contained gold and silver, the high costs of production discouraged mining. The existence of local mills did not lower these costs sufficiently to justify their operation. Additionally, World War II limited the availability of financing, materials, and laborers. Only 377 tons of ore were produced in 1942.

1943 - 1982: The Quiet Decades

However, hope springs eternal for a Midas miner. A few mines and prospectors continued operations sporadically over the 50 years after the major mines and mills closed.

The 1942 *Minerals Yearbook* indicated that the Miners Gold operated for four months and was one of four producing mines in the district that year. In 1943, only one mine reported its operation, producing a mere 12 tons of ore. According to Jack Murdock, the only employment available in the area during the war years was the dismantling of the Gold and Silver Circle Company's mill. After the 1943 report, the district was silent until 1947.

According to the *Nevada Mines 1947*, the Miners Gold was still operating that year. The publication described the operation:

> This is an old property with two thousand feet of underground workings. They [operators of the Miners Gold Mining Company] have reconditioned the old workings and sunk a new shaft. Equipment consists of bin, compressor, air hoist, change room, compressor room, and all necessary mining equipment. They are planning to build a mill of 25 to 30-ton capacity.

The Miners Gold as it appeared in 1958 (Photo courtesy of Phil Gemmill, Friends of Midas Collection)

Herb Lewis and Bill Gemmill (on the right) hard at work in 1958 (Photo courtesy of Phil Gemmill, Friends of Midas Collection)

The entrance to the Missing Link Mine in 1958. (Photo courtesy of Phil Gemmill, Friends of Midas Collection)

Mining man Bill Gemmill in 1959. (Photo courtesy of Phil Gemmill, Friends of Midas Collection)

Production, mostly from recovery work at the mine dumps and some prospecting, continued into the 1950s. The Reno newspapers announced, in August 1950, that uranium had been discovered near Midas. In 1957, the district reported the production of only one ounce of gold and one ounce of silver, yielding a total of $36. Other reports revealed that mining in the area increased somewhat after 1957. Operating mines were reported every year from 1957 through 1965, although the tons of ore produced and their yields were not always indicated. The *Minerals Yearbook* noted that development and rehabilitation work was done at the Miners Gold and Dixie Gold mines in 1959. In 1961, the Nevada Gold mine produced 75 ounces of gold and over 200 ounces of silver. Two years later, a reporter from *The Humboldt Star and Battle Mountain Scout* observed "a new mining outfit drilling on the side of the mountain, within sight of town. It appeared to be quite a project, in view of the number of pieces of equipment being used." In 1965, the Dixie Gold reported shipping gold ore to a Utah smelter. Midas Gold Company, owned by Victor K. Sorbe and L. Don Lott, paid Andy Jones for work he performed that year at the Dixie Mine and Mill.

In every year from 1965 to 1970, however, at least one Midas mine reported ceasing operations. After 1970, the state and national reports were mostly silent for over 25 years. But mining in the district did not stop completely. In 1982, state geologists reported that the Ripsaw open-pit gold mine was in operation. Located one-half mile northeast of Midas, it employed four people. Additionally, several major exploration projects were conducted throughout the 1980s, but these activities did not result in any discoveries worthy of being worked.

The portal to the Ken Snyder Mine was designed to resemble the portal to the Sutro Tunnel in Sutro, Nevada. (Photo courtesy of Newmont Mining Corporation)

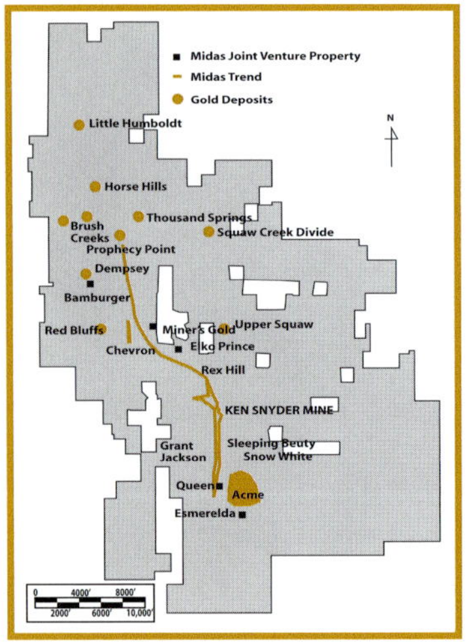

Franco-Nevada Mining Company featured its Midas mining operation in its 1999 annual report.

1991-2006: Mining Resumes

Then, in August 1994, geologist Dr. Ken Snyder made a discovery of which part-time prospector James McDuffy could only dream. Through persistent drilling and high-tech geological modeling, Snyder discovered a deposit estimated to hold four million ounces of gold. *The Humboldt Sun* reported that this was a deposit that the first miners had simply missed. One group of early miners had stopped drilling 50 feet from the deposit; a later group started their drill work too high. Andre Douchane, Vice President of Operations for Midas Joint Venture, explained to the Winnemucca newspaper:

> Although the outcrop has been obvious for years, it doesn't have any gold in it for about 500 feet and, unless you put an angle hole or a deep vertical hole in, you weren't going to find it.

Snyder did find it, which ultimately proved to be the richest gold discovery in Nevada since Goldfield in 1902.

Snyder's work in the district dated to 1991 when he had begun to acquire properties in the district on behalf of Midas Joint Venture, which was an equal partnership between Toronto-based, publicly-traded mining corporations Euro-Nevada and Franco-Nevada. By 1994, Midas Joint Venture had acquired approximately 95 percent of the district. Exploratory drilling began, and the steeply dipping, narrow, high-grade gold and silver vein was soon found.

One of the obstacles to earlier mine production in the Gold Circle Mining District was the haphazard land ownership pattern in the area. Over a period of three years, Midas Joint Venture quietly bought land and claims, dealing with 36 different groups. By 1998, with the purchase of 325 acres from Romarco Minerals, Inc., the company had acquired the vast majority of the historic Gold Circle Mining District. It was the sole owner of 26,000 acres (40 square miles), only five percent of which had been explored by the time the 21st century began.

In January 1997, Midas Joint Venture announced the

beginning of a process that would result in a 500-ton-per-day mining and milling facility. The mine portal, designed after the historic Sutro Tunnel Portal, was built in March of that year. Excavation began of the mine, which would be a 19,000 foot decline corkscrewing from the portal to a depth of 1,800 feet, according to *The Humboldt Sun*.

Franco-Nevada Corporation's 1999 annual report, *New Engines of Growth*, outlined the project's timeline. Permitting was completed in February of 1998. The mine was the only one in Nevada fully and only permitted by state authorities because of its placement on private land. Early in the mine's development, company officials consulted with Midas residents and others concerned about potential impacts. After a meeting with Native American representatives, the mill's planned location was changed to avoid placing the facility on sacred land. Cooperating with community representatives, the company worked to mitigate the effects of a large, modern mining operation on the small, quiet town. All of these activities contributed to a speedy conclusion to the permitting process.

Construction of the mill and ancillary facilities had begun in January of 1998. In July, Midas Joint Venture contracted with Dynatec Corporation to operate the complex and provide all of the necessary personnel and underground mining equipment. Three months later, additional historic claims were purchased from Romarco Minerals, Inc., with Euro and Franco stock worth $21 million, according to the *Elko Daily Free Press*. Drilling had determined that the acquired area held a 460,000-ounce gold resource. On November 21, 1998, the mill, which was permitted to process 700 tons of ore per day, started for the first time. The planned structure of the mill had been described by *The Humboldt Sun* in 1997:

> Ore will be crushed in jaw and cone crushers and ground in a conventional ball mill to minus-200 mesh, then processed in a whole-ore leach with Merrill-Crow gold precipitation. Cyanide in tailings will be destroyed with the Inco sulfur dioxide process.

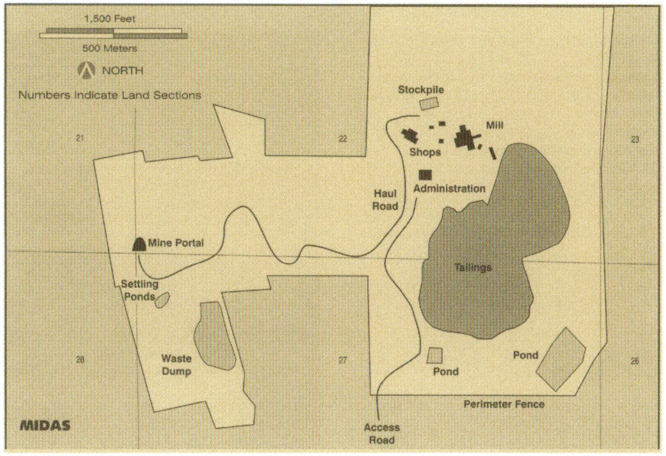

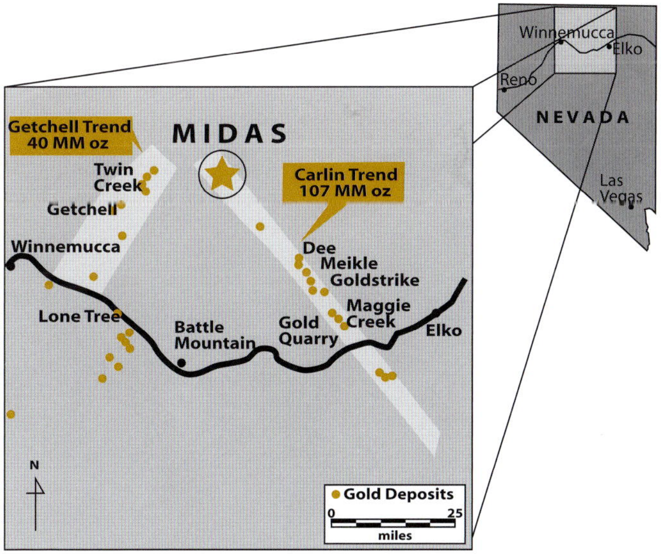

Newmont Mining's Midas Operation. (Maps courtesy of Newmont Mining Corporation)

Midas processing facilities around 2003 (Photo courtesy of Newmont Mining Corporation)

On December 9, 1998, the first bars were poured, consisting of 291 ounces of gold and 2,915 ounces of silver. Completed four months early and within the $85 million budget, the mine and mill were declared commercial on January 26, 1999.

Franco's 1999 annual report, understandably enthusiastic, outlined the mine's physical and financial attributes:

> It is a near vertical, very continuous vein type deposit. It is very dry by normal mining standards. It is shallow, all within 1,500 feet of surface and easily accessed through a decline. No public lands are disturbed, and the waste rock does not generate acid. It will be one of the lowest cost gold mines in the world.

Thus, Midas Joint Venture began what was expected to be 10 years of mining activity in the Gold Circle Mining District.

However, on May 31, 2001, Normandy Mining Limited, Australia's largest and the world's sixth largest gold producer at that time, announced the completion of its transaction with Franco-Nevada, which included acquisition of the Ken Snyder Mine. Midas Joint Venture ceased to exist, and Dynatec was no longer the operator. Although "Ken Snyder Mine" remains on the portal, it is now more commonly referred to as the "Midas Mine."

Almost a year later, on January 31, 2002, Newmont Mining

Newmont's Midas Mine and Mill (Photo courtesy of Newmont Mining Corporation)

Corporation announced the expected acquisition of both Normandy and Franco-Nevada for $4.41 billion in stock and cash. Completed in mid-February, the deals helped establish Newmont as the largest gold producer in the world at the time. Headquartered in Denver, Colorado, Newmont Mining is a publicly traded company with mining operations in nine countries.

Currently, Newmont operates the Midas Mine with about 226 employees. According to the company's website, the Midas Mine "sold approximately 219,000 ounces of gold and more than 2.6 million ounces of silver in 2003, representing approximately 8.4% of Nevada's total gold sales."

The U.S. Geological Survey ranked Nevada as second among the 50 states in silver production in 2002 and 2003. Those same years, the Midas Mine was Nevada's second biggest producer of silver behind the Rochester Mine near Lovelock.

With an expected mine life of another five years, 950 tons of ore are currently mined and milled daily, which provides an interesting contrast to historic production. Had the ore mined between 1907 and 1965 been produced at this same rate, it would have taken slightly more than 309 days to mine the 294,172 tons. In less than one year, the new Midas Mine produces more gold and silver ore than the historic Gold Circle Mining District did in 57 years.

From 1907 through 1965, all of the mines of the Gold Circle Mining District produced, in total, 129,477 ounces of gold and 1,656,184 ounces of silver with a value of just under $4 million. In the first full year of production alone, the Ken Snyder Mine exceeded that yield by producing 182,345 ounces of gold and 1,860,701 ounces of silver with a value of $25.3 million. Nevada's Division of Minerals reported that the total production of the Midas Mine, from 1998 through 2006, was over 1.4 million ounces of gold and 16.5 ounces of silver.

John W. Cody's sons, Bill and John, with their burros Rube and Ruby (Photo courtesy of Lela Halsell Hempton, Friends of Midas Collection)

The Dangers of Mining

Underground mining is hazardous work, and mills can be dangerous places. However, despite the large number of people who have worked in the mines around Midas during the past 100 years, only five were killed in mining-related accidents. None were buried in the Midas Cemetery.

The district did not exist for many years before the danger inherent in its main industry was clearly illustrated by the first mining casualty. In 1910, Nevada's Mining Inspector reported that Cyrus Lee Gilson, General Superintendent at the Rex, was killed "almost instantly" by mill machinery on March 23. Evidently, he had become entangled in a belt while trying to unwind it. When the Battle Mountain paper recounted the accident, it noted that the 32-year-old Gilson was considered an expert millman and had recently worked in Mexico. His father was famous for discovering the mineral-rubber, Gilsonite.

On November 22, 1921, the second mining casualty occurred in the district. According to the annual report from Nevada's Inspector of Mines, Bernard Daxon Hayes, aged 39 years, was killed by an accident in the Elko Prince Mine. Foreman of the mine, Hayes fell 150 feet into the mine as a result of old timbers collapsing beneath him. Although he was retrieved immediately and given medical attention, he died within the hour. Hayes had been in Midas for over four years. Buried in Oakland, California, he left wife, Mary, and a son, Frederick, who had been born in Midas in 1918.

The hazards of mining were graphically demonstrated again in 1927 with the death of long-time Midas miner, John W. Cody. According to the Battle Mountain *Scout* and the state's annual mining report, Cody was killed in the Grant shaft belonging to the Gold Circle Consolidated Mining Company on December 2, 1927. Cody was a hoist man on the night shift. Apparently, one of the cars had become stuck and, while trying to release it, Cody fell into the car, which then plunged 165 feet into the mine. Somewhere during the fall, Cody's head struck a metal object and was, noted the Battle Mountain *Scout*, "split open from the base around the top to his nose." He was probably dead before he reached the bottom of the mine. Cody, 59 years of age, was married with children and a 20-year resident of Midas. He and Mike Paisley had discovered the Missing Link vein early in the district's development period. Cody, who also owned a restaurant in Midas, was buried in Winnemucca. He was survived by his wife, Mamie, and sons, Bill and John.

Almost nine years later, on June 5, 1936, George Anderson, 20 years old, was severely injured shortly after starting his night shift in the mill of the Gold and Silver Circle. Apparently, he failed to remove the handles on a motor that revolved at a high rate of speed. The motor started, and, according to the Mine Inspector's report, Anderson "was flayed by the projecting pieces of steel." The Battle Mountain *Scout* related that Ralph Duncan, the night mill man, shut off the machinery and tried to rescue Anderson. Duncan suffered a broken finger and chest bruises in the attempt. Beaten severely and with several broken bones, the still-conscious Anderson was taken to Winnemucca. But, Anderson, described by the *Scout* as the "brilliant young engineering student at the University of Nevada," died en route. The newspaper noted that this fatality was the first to occur at the mill.

Over a half-century passed without another fatality. Then on October 23, 1999, Jean Luc Doucet, 35, of New Brunswick, Canada, was killed in the Ken Snyder Mine. A longhole driller, Doucet died at the 5,300 foot level of the underground mine when a 3,100 pound boulder fell and pinned him. The mine's operator, Dynatec, was cited by federal mine safety officials as a result, according to the *Reno Evening Gazette*.

Modern mining companies endeavor to ensure that their employees are as safe as possible. The Ken Snyder Mine was originally built with a ventilation raise equipped with a hoist to provide an additional escape route, should one ever be needed. In 2004, Newmont Mining reported that its Midas operation experienced a significant drop in the number of work-related illnesses and injuries. The same year, the Midas Mine was the second place winner of the Nevada Mining Association's award for safety in underground mining operations. Midas Mine Supervisor Todd Melonas was recognized by the Association in 2005 for exhibiting outstanding dedication to safety.

GOLD CIRCLE MINING DISTRICT
ANNUAL PRODUCTION: 1908-1965

Year	No. of Producers	Number of Tons	Ounces of Gold	Ounces of Silver	Other (Pounds)	Total Value
1908	6	245	1,769	1,172		$ 37,182
1909	4	1,815	4,288	3,717		$ 90,568
1910	12	1,042	1,198	1,152		$ 25,389
1911	7	7,284	2,842	3,668	13 copper	$ 59,784
1912	14	3,137	1,793	5,173		$ 40,235
1913	11	2,626	1,642	1,333		$ 34,746
1914	9	3,362	3,008	11,725	23 copper	$ 68,677
1915	10	2,946	2,040	13,704		$ 49,116
1916	11	23,947	11,787	157,207		$ 347,107
1917	16	19,523	11,716	182,844		$ 392,861
1918	6	18,692	13,961	167,684		$ 456,275
1919	6	15,586	9,462	132,079		$ 343,532
1920	5	14,000	5,054	94,109		$ 207,061
1921	9	12,733	5,885	82,967		$ 204,447
1922	8	NR	2,269	28,748		NR
1923	5	165	105	85		NR
1924	3	183	461	8,331		$ 15,133
1925	5	61	33	41		$ 708
1926	2	NR	22	17		NR
1927	NR	NR	6,740	94,871		NR
1928	5	26,170	8,213	149,026		$ 256,950
1929	4	13,913	5,809	82,588		$ 164,097
1930	3	552	378	4,216		$ 9,447
1931	3	4,841	2,681	31,798		$ 64,634
1932	7	7,231	3,787	66,568		$ 97,065
1933	8	2,968	2,587	36,540		$ 66,265

Year	No. of Producers	Number of Tons	Ounces of Gold	Ounces of Silver	Other (Pounds)	Total Value
1934	9	5,330	2,728	42,765		$ 123,003
1935	11	5,492	1,016	12,567		$ 44,594
1936	10	17,305	4,238	58,541		$ 193,664
1937	4	17,182	2,808	27,073		$ 119,221
1938	6	25,856	2,442	54,717	800 lead	$ 120,879
1939	3	29,625	2,916	55,007		$ 139,398
1940	6	6,767	1,766	14,641		$ 72,221
1941	9	2,857	1,461	21,514	300 copper	$ 66,174
1942	4	377	248	2,132		$ 10,196
1943	1	12	51	578		$ 2,196
1944-46	NR	NR	NR	NR		NR
1947	2	225	3	579	300 copper 18,000 lead	$ 3,284
1948	3	31	23	216	100 lead	$ 1,018
1949	2	11	4	6		$ 145
1950	5	47	98	3,648		$ 6,732
1951	3	33	31	497	100 lead	$ 1,559
1952-56	NR	NR	NR	NR		NR
1957	1	NR	1	1		$ 36
1958	NR	NR	NR	NR		NR
1959	NR	NR	12	14		NR
1960	1	NR	10	19		NR
1961	NR	NR	75	203	200 zinc	NR
1962-64	NR	NR	NR	NR		NR
1965	NR	NR	16	103		NR
Totals		294,172	129,477	1,656,184	19,836	$ 3,935,599

NR = Not reported

Sources: United States Bureau of Mines and Nevada State Inspector of Mines

JONATHAN PAYNE, Resident Agent
Principal Office, RENO, NEVADA
INCORPORATED UNDER THE LAWS OF
NEVADA
TREASURY STOCK
Shares -1,000-
Eastern Star Mining Company
CAPITAL STOCK, $1,000,000
1,000,000 SHARES
PAR VALUE, $1.00 EACH
FULLY PAID AND NON-ASSESSABLE
This Certifies that Mrs. John Reimmiller is the owner of
One thousand Shares of the Capital Stock of
EASTERN STAR MINING COMPANY
transferable only on the books of the Corporation by the holder hereof in person or by Attorney upon surrender of this Certificate properly endorsed.
In Witness Whereof, the said Corporation has caused this Certificate to be signed by its duly authorized officers and to be sealed with the Seal of the Corporation
this 29th day of November, A.D. 19 12.
PRESIDENT
SECRETARY
SHARES $1.00 EACH

MIDAS MINE PRODUCTION, 1998-2005

Year	Operator	Number of Employees: Company	Contract	Ounces of Gold	Ounces of Silver
1998	Dynatec Mining Corp.	13	95	4,357	55,329
1999	Dynatec Mining Corp.	170	11	182,345	1,860,701
2000	Dynatec Mining Corp.	229	5	197,800	1,941,989
2001	Normandy Midas Operation, Inc.	159	74	198,518	2,393,246
2002	Newmont Mining Corp.	170		218,966	2,647,374
2003	Newmont Mining Corp.	162		232,949	2,870,164
2004	Newmont Mining Corp.	167		219,778	2,471,135
2005	Newmont Mining Corp.	217		167,297	2,166,396
Totals				1,562,010	17,406,334

Source: Doug Driesner and Alan Coyner, The Nevada Division of Minerals. *Major Mines of Nevada: Mineral Industries in Nevada's Economy.* Nevada Bureau of Mines and Geology, 1999-2006.

In April 1908, these tents were located in the area where the Elges place now is. (Photo courtesy of the Nevada Historical Society)

THE TOWNSHIP OF MIDAS

The town of Gold Circle began with the mining district for which it was named and was located in a treeless, narrow canyon within walking distance of the major mines. Contrary to a popular myth, the mines did not encircle the town. At best, they formed a semi-circle north, east, and south and are physically separated from the town by low-rising hills. Within a year after McDuffy's strike in 1907, the township was officially designated by Elko County, and the name was changed to Midas. The new town (known as Gold Circle or Midas, but never Rosebud) had a federally-sanctioned post office, a justice court, an election precinct, and other trappings of an organized community. Although the government-sponsored districts were all dissolved by 1972, the modern needs of Midas have been met by community-driven organizations.

Establishing the Township

On October 31, 1907, the Battle Mountain newspaper reported that lots in the new town were selling briskly for $50 to $250 each. At some point during that fall, Patrick A. Leamy organized the Gold Circle Townsite Company, which became the primary source for property. A short two months later, the Tuscarora paper noted that frame structures were being built, and the cost of a lot had doubled. On January 30, 1908, the Battle Mountain paper exclaimed that the "town . . . is springing up as if by magic. Transactions in real estate are being constantly made and one day 27 lots changed hands in an hour's time." *The Gold Circle News* printed a weekly list of real estate

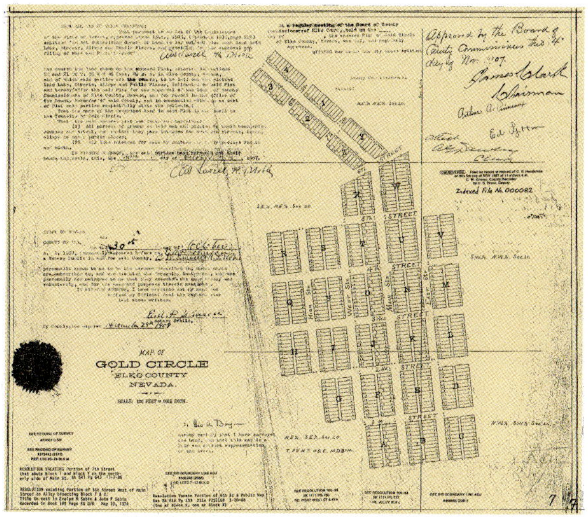

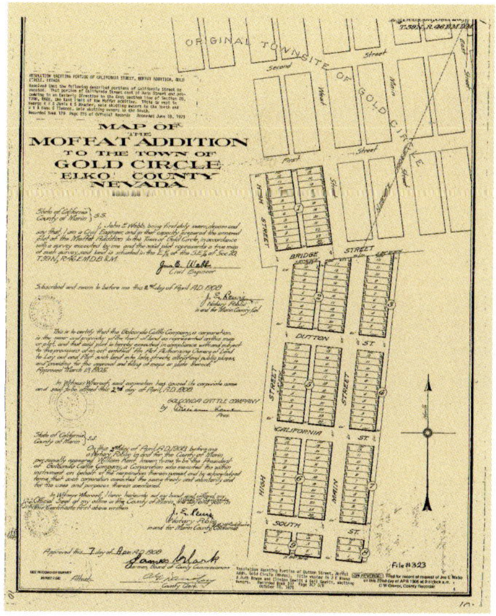

Plat maps for the Gold Circle Townsite and its first addition (Courtesy of the Elko County Recorder)

Gold Circle, Nevada, around 1907 (Photo courtesy of Jerry Fenwick)

The newly-christened and growing town of Midas on May 1, 1908. The large tent in the foreground is probably a hotel. (Photo courtesy of Jerry Fenwick)

It was not uncommon for people to gather in front of a local establishment (note the barber pole by this building), particularly if the mail was expected. The post office was the next building to the right, just out of the picture. The building on the hill, above the awning, was the teacherage. (Photo courtesy of Edna Wilkerson Timmons, Friends of Midas Collection)

transfers; for the week ending June 13, 1908, 18 lots were sold for prices ranging from $100 to $650 each.

But the building of the town was not without complications. In April 1908, *The Gold Circle Miner* reprinted a story from *The Carson News* that John G. Taylor, a prominent Nevada sheepman, was claiming ownership of "the entire townsite district upon which the new mining camp of Gold Circle is located." Many of the details about this issue and its resolution are not known, but documents on file with the Elko County Recorder provide some interesting information. On April 11, 1908, Taylor sold approximately 160 acres, which appear to contain the townsite, to Leamy. Two days later, Leamy sold most of this land to the Gold Circle Townsite Company of which he was the principal agent. The Townsite Company transferred only the surface rights to the buyers of its lots, reserving the mineral rights for the company.

Taylor's claim was probably settled with the purchase by Leamy, but a boundary dispute appears to have arisen later. Less than a year after the original sale, Taylor, the Gold Circle Townsite Company, and the Golconda Cattle Company filed a formal agreement with Elko County, specifying the townsite's location after a formal survey was completed.

Although the townsite had been laid out in the canyon and lots sold, the town of Gold Circle was not officially established until the Elko County Commission approved the town's plat map on November 4, 1907. Other official actions concerning the town quickly followed. The post office was approved by the U.S. Postal Department on November 16, 1907. The justice and election precincts were organized by Elko County in March and July, respectively, of 1908.

The official name of the post office and precincts, however, was not "Gold Circle." When the residents of the town, represented by Leamy, applied to the Federal Government for a post office, they were not granted their first choice for its name. Obviously, their choice was "Gold Circle," but the administration refused to approve yet another post office in Nevada with the prefix "Gold." In 1907, several "gold" post offices were operating in Nevada: Goldbar, Gold Butte, Gold Center, Gold Creek, Goldyke, Goldfield, and Gold Hill. So the name "Midas," which had been used by the post office at Ione in Nye County until 1903, was assigned to the town. The Elko County Commission followed suit, naming the election and judicial precincts "Midas."

The name may not have been assigned to the town arbitrarily by the Post Office Department. In fact, the name may have been suggested by the townspeople. Although the post office application did not ask the applicant to indicate more than one choice for a name, Leamy may have offered four options. The original application lists, in apparently the same handwriting, four possible names for the post office: Gold Circle, Midas, Mint, and Porphyry. (Porphyry is a type of rock found throughout the district. The word was used often by *The Gold Circle News* when describing the progress on various claims.) It appears that the Post Office Department may have simply chosen the town's second proposal for its name.

Further, it seems that the name was not suggested arbitrarily by the residents. At that time, one of the major mines in the district was the Midas. Located by P.W. Dargin, R.E. Ashbrook, Richard Bamberger, and Gus Guethlein in the summer of 1907, the Midas group of claims was often mentioned as one of the district's first and richest properties in early newspaper articles about the new area. The Carson City *Daily Appeal* exclaimed that "one of the greatest strikes in the history of the state" was made on the property and included "a two-inch streak of pure gold running through the ledge." Throughout the history of the district, the Midas was named as one of the region's best producers. Additionally, Bamberger was often credited with the first significant find that led to the establishment of the district. It seems reasonable that a mine as important as this one was in 1907 supplied an alternative name for the town.

The truth behind the name choice may never be known. After all, someone at the Post Office Department, not Leamy, may have written in the name choices. The mining district's name was not officially changed, but "Midas" and "Gold Circle" were—and still are—freely interchanged as the names of the town and district. The residents continued to refer to their town as Gold Circle for many years, but the name Midas was not entirely disagreeable to them. In fact, the implication of the new name was used to its full advantage. An advertisement in *The Gold Circle Miner* for Midas had the headline, "Touch it and it turns to wealth."

At the Elko County Commission meeting on February 10, 1908, the residents of Midas presented a petition asking for the creation of a township and the appointment of officers. Also in February, the miners of the district met to adopt "the rules and regulations by which the district will be governed," reported the *Battle Mountain Herald and Central Nevadan*. The article stated that:

> A committee of ten has been appointed [to] arbitrate differences between miners . . . This committee will also look after habitual claim jumpers who may try to dispossess any of the prospectors through technicalities in surveying and monumenting claims.

A month later, the county commission created the Township of Midas and specified its borders. According to the Elko *Daily Free Press*, the township was huge: the western border was the dividing line between Elko and Humboldt Counties, and the southern was the line between Lander, Eureka, and Elko Counties (the other two borders were survey lines).

Competing Towns

With the first rush to the district, there was no guarantee that Midas would become the premier town in the area. It certainly was not the first location where people settled. In fact, several prospectors probably came from Dutton, a camp that was located on Rock Creek, not far from the new find. Dutton's application for a post office noted that 25 people would be provided mail services and was signed by Charles D. Dorsey, "proposed postmaster." The application was approved on June 11, 1907; most likely, the camp had been formed several weeks, perhaps months, earlier. Dutton's post office operated until September 30, 1913, when its mail service was transferred to Midas.

As with the development of any new district, the establishment of a town was second only to the mines. The town of Gold Circle appeared almost immediately. But it was not the only one. Victoria and Summit also vied for the honor—and anticipated wealth—that would come with anchoring the district.

Victoria, however, was not much of a contender. It appears that this town may have been no more than an idea in the minds of E.A.P. (Edgar) Johnson, R.T. (Dick) Noble, R.J. Pemberthy, Charles A. Stoneham, and Samuel Dowling. On the 1907 location notices they filed with the Elko County Recorder, they noted that their claims were near the "Victoria Townsite" in the Golden Circle Mining District. But their plans appear to have fizzled as the town was not mentioned again.

Summit was a more serious venture. The town was located on the flats about a mile south and west from the town of Gold Circle. More developed than Victoria, Summit did not hesitate to declare publicly that it should be the premier town in the district. A large advertisement in the April 11, 1908, issue of *The Gold Circle Miner* proclaimed Summit "the logical townsite of the Gold Circle District" with its "natural drainage" and "the best pure water system." "This townsite," it went on to claim, "is the nearest one to the big mines. It is the only one where

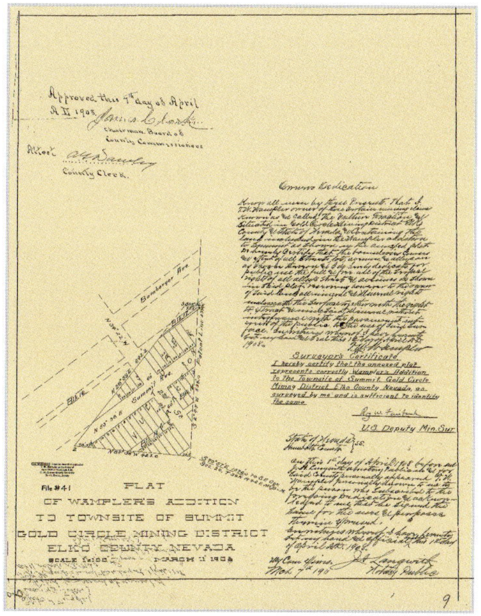

Wamplers Addition to Summit (Courtesy of the Elko County Recorder)

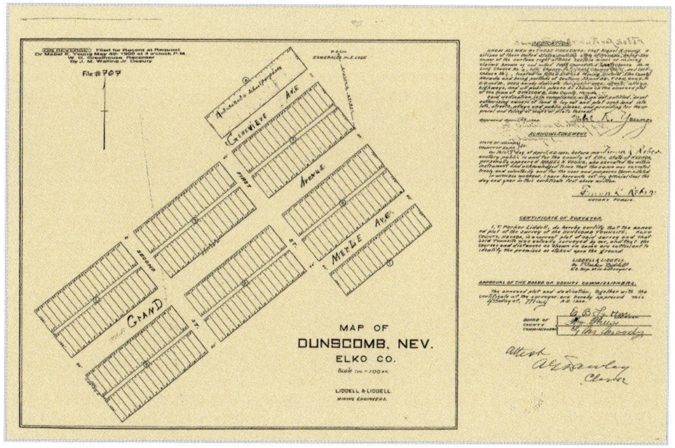

It is not known if any of these lots in Dunscomb were sold. (Courtesy of the Elko County Recorder)

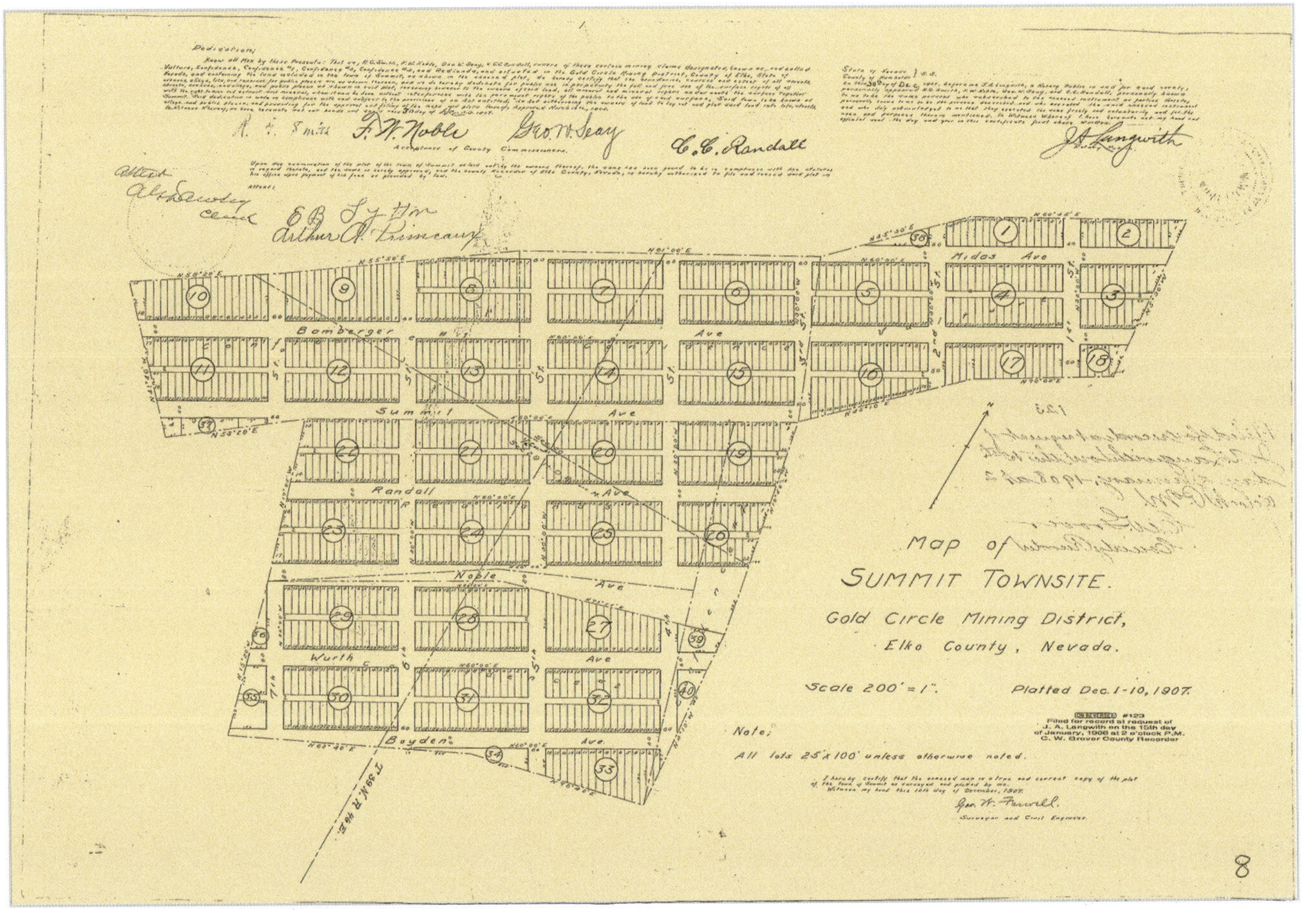

Summit's plat map expected much larger development of the town than actually happened. (Courtesy of the Elko County Recorder)

water will be piped into your house." The ad also mentioned the improvements and building going on within the town and reminded the reader that "when the sun shines, it shines on Summit." The Battle Mountain paper reported that Summit had a lodging house, a butcher shop, and (of course!) a saloon. Additionally, the Gold Circle paper advertised the locations in Summit of a restaurant and a branch of the Golconda Mercantile and Banking Company.

The Elko County Commissioners approved a plat map for the Summit Townsite on January 6, 1908, and for the Wampler's addition to the town on March 11, 1908. The town was also granted a post office with John Slavin serving as postmaster. The Summit Townsite Company worked vigorously to attract new residents. A large advertisement in the local paper listed the following officers for the company: Clayton Gannett, President; F.W. Noble, Vice-President; and George F. Busch, Secretary.

Their best efforts could not keep Summit alive. The post office operated for a mere 17 months, from February 8, 1908, to July 31, 1909, when this settlement was finally and completely eclipsed by the town of Gold Circle, which soon became more commonly known as Midas.

Other towns attempted to come to life in the area. On April 20, 1909, Midas residents Al Geabbeart and Edward Dunscomb, M.D., agreed to transfer the surface and water rights of the Lucky Chance group of claims to Dr. Mabel Young, the town's dentist. The contract stated that the land was transferred for the expressed "purpose of surveying and laying out . . . 'The Dunscomb Townsite.'" Dr. Young, President of the Dunscomb Townsite Company, was granted sole authority to develop and sell lots in the townsite. In return, Dr. Dunscomb and Geabbeart would receive one-half of the proceeds from the sale of lots in the site. Soon after, the Elko County Board of Commissioners approved the plat map for Dunscomb, which was to be built near the Esmeralda mine approximately two miles from Midas, and the Dunscomb Townsite Company was authorized to sell lots. The Battle Mountain paper noted that lots would be for sale after June 1. However, the plans for the town appear to have been cancelled shortly thereafter.

In October 1912, the Elko newspaper mentioned that the Eastern Star Mine was doing so well that the small town of Star was planned. According to the reporter, "the distance between the future town of Star and the present town of Midas precludes the probability of anything more than the spirit of friendly rivalry." The excitement appears to have been short-lived as the town was not built.

The Post Office

By March 1908, Midas was receiving mail weekly through Tuscarora. In September, mail service increased to six days a week. In the 1930s, mail arrived on the truck driven by Irene True between Winnemucca and Midas three days each week, and Noreen Wilkerson Murdock remembered being required to ski out to meet the mail truck when it could not manage to get through the snow to town. However, by the town's fourth decade, the population had declined to a point where a post office could no longer be supported. On September 16, 1942, the post office was closed permanently. In the 34 years of its existence, seven postmasters had served the community; one held the assignment for over 14 years.

Despite the closure of the Midas post office, regular mail delivery continued to be a part of town life. After the Federal Government closed the town's post office, mail traveled to and from Midas via the Golconda post office. The Elko *Daily Free Press* observed that the Post Office Department planned to discontinue this mail route on June 30, 1943. However, that plan was dropped when U.S. Senator James G. Scrugham intervened after receiving numerous protests from "interested citizens of that area." The route was continued, and mail was delivered in Midas three times a week most of the year. Delivery was available only two days a week during the first three months of each year because of the harsh weather and difficult road conditions.

Midas continues to receive its mail from Golconda, but more frequently than 50 years ago. Mail service is provided daily, six days a week, by courier from the Golconda Post Office because Midas has neither its own post office nor zip code.

This building served as the Midas post office for many years. (Photo courtesy of Edna Wilkerson Timmons, Friends of Midas Collection)

Midas Postmasters, 1907-1942

Name	**Appointment Date**
Patrick A. Leamy	November 16, 1907
Lynn B. Yadem	October 30, 1908
Mabel K. Young	July 16, 1909
Margaret Brady	August 14, 1911
Charles S. Warren	September 7, 1916
Miriam Stenson	October 18, 1919
Isabel Lyon	April 30, 1928
Post office closed	September 16, 1942

The Justice Court

When the township of Midas was established, the Elko County Commission also appointed Leamy as Justice of the Peace. Leamy, however, declined. On April 6, 1908, E.C. Rickard was appointed because his supporting petition from the citizens of Midas had more signatures than did the petition for J.W. Matthews. A month earlier, the Elko County Sheriff had appointed Guy Harbin as Deputy Sheriff for Midas.

From 1908 through the election of 1950, local voters elected their neighbors to fill the offices of Constable and Justice of the Peace. Neither job was terribly time-consuming. The Constable was typically concerned about inebriates, and the Justice of the Peace performed weddings. However, the justice court did record its official activities, and its civil and criminal docket was maintained at the Justice Court on Main Street in Midas. Many years later, Maxine Sweeney recalled that, after her last term expired as Justice of the Peace, she delivered the Midas Justice Court docket to the father of former Elko County Manager George Boucher. The younger Boucher traced the docket to an old house in Elko that had been burned in an exercise for the local fire department, destroying all of the records it had housed.

The last year that both local offices were partisan was in 1908 when Democrats James W. Matthews and John Boyle were elected Justice of the Peace and Constable, respectively. After that year, the candidates for Constable often declared a party affiliation, but the Justice of the Peace became a nonpartisan office.

The Elko *Daily Free Press* reported that the 1912 winners for Midas township officers included George A. Strough, Justice of the Peace, and J.M. Wyatt, Constable. The candidates ran unopposed, except Strough, who defeated James M. Ervin by a margin of 63 votes.

Records at the Nevada Historical Society indicate that, in 1916, J.M. Wyatt was elected Constable over George Rosenfelt. John Bake was chosen in 1918, but the records note that it is unknown if he had an opponent or who was elected in 1920. It appears that George Parkinson was elected in 1922 because the *Nevada State Journal* reported, before the 1924 election, that the Elko County Commission appointed Jack Cody to finish the term because Parkinson had moved from the district.

In 1928, the *Battle Mountain Scout* noted that Jess Snyder was re-elected Midas Constable by three votes. In April 1931, the paper reported that Mike J. Paisley was appointed Midas Constable by the Elko County Commission until another election was held. Paisley replaced Thomas Nekirk who had died after serving as constable for an unspecified period of time. Three years later, the paper mentioned that J.W. Phelps was the constable. At the 1934 election, E.D. Brenker was elected constable "by a large majority," but he must not have completed his term as the County Commission appointed Al Fox to the vacant position in October 1935. Records at the Nevada Historical Society related that Fox was re-elected in 1935 by 44 votes over James C. Wilson; Brenker defeated H.E. Lauritzen a year earlier by 59 votes; and Snyder had bested M. Alberson, 77 to 74.

In 1938, Al C. Fox was again elected constable. Lou Wilkerson, a Democrat, was elected in 1942 and paid a monthly salary of around $73. Two years later, Midas voters chose the incumbent Wilkerson over Joe Horvath (the same "Hungry Joe" arrested for a 1937 shooting and probably the last to be incarcerated in the

Constable Lou Wilkerson. (Photo courtesy of Edna Wilkerson Timmons, Friends of Midas Collection)

Midas Jail). Both men were Democrats. In 1946, Wilkerson was unopposed. In 1948, W.S. Allen beat Jack Ames for the office of Constable by 17 votes. In 1950, no one ran for the office of Midas Constable, possibly indicating that the office had been abolished.

The Justice of the Peace that most former residents remember was Frances Marion "Frank" Macy. He was first appointed to the office on or before May 5, 1912, after Judge Strough resigned. The two traded the office again. A few months later, Strough was re-elected but, apparently, resigned again by the end of 1914. Macy was re-appointed on January 5, 1915, and then elected to the post in 1916 and 1918.

In 1922, however, Richard T. "Dick" Noble was elected Justice of the Peace over Sam Adler by a vote of 47 to 26. The records are silent until 1928, when Frank Macy ran unopposed and collected 137 votes. Six years later, Macy defeated Jean Baker by 60 votes. In 1936, 1938, and 1942, Macy was unopposed and again re-elected, serving in this office until his death on March 2, 1943. The "Midas News Notes" rarely mentioned Judge Macy in an official capacity, other than his performing a wedding or two each year.

After Macy's death, the Elko County Commission appointed long-time Midas resident Luella Pendergast to the post. In 1944, however, Maxine Sweeney defeated Pendergast for Justice of the Peace, 21 to 12. Sweeney was re-elected in 1948 and 1950 unopposed. She was the town's final Justice of the Peace.

The Election Precinct

Before the creation of mail precincts, which is the manner by which Midas residents currently cast their local, state, and national votes, Election Day was an important event in town life.

The Gold Circle News of September 26, 1908, was devoted to preparing the residents of Midas for the coming elections, which would include voting for township officers for the first time. Evidently, the citizens had a strong interest in politics. By October 1908, 193 men had registered to vote in the Midas precinct, ranking the precinct third in Elko County for the number of voters registered. First was Elko precinct (525 registered), then Carlin (with 212 voters).

According to the paper—perhaps with the intent to make it so—the voters of Midas were predominantly Democrats. Editor Frank L. Reber encouraged his *News* readers to vote for William Jennings Bryan for United States President and to contribute $1 each to his campaign. Enlisting a little editorial license, Reber exclaimed:

> Wait until election night and you will hear a ring as pure and strong from Gold Circle for Democracy as its . . . gold will ac claim the world and enforce the strength of Gold Circle as the greatest gold producer of America.

Although Reber's claims about the future of the district were overstated, it does appear that most of the voters in the district registered with the Democratic Party. That year, Midas sent more men to the county Democratic convention than did most of the other towns in Elko County. Before the election, Reber listed 63 men as members of the "Gold Circle Democracy," which he called "a fighting battalion, yet a generous and sympathetic combination, that is considerate of the rights and fellowship of their opponents." He printed a positive, almost angelic, phrase about each man listed, including one about whom he said "we don't know [him] exactly, but cannot believe anything else but that he is a true man."

The results of the general election, held November 3, 1908,

William Jennings Bryan is framed by the issue of the Gold Circle News that predicted incorrectly that he would win the U.S. Presidency. (Newspaper courtesy of the Nevada Historical Society)

Francis G. Newlands (Courtesy of the Special Collections Department, Getchell Library, University of Nevada, Reno)

reflected the confidence of Midas voters in the Democrats. Approximately 130 men cast their ballots at the official polling place for the Midas precinct, the Canyon Hotel. Most of these voters chose the Democratic candidates for President (Bryan), U.S. Senator (Francis G. Newlands), and U.S. Congressman (George A. Bartlett). Newlands and Bartlett were both statewide winners. Republican candidates for these offices came in second among Midas voters, as they did in the rest of the state, and Socialist candidates, third.

Midas voters also participated in county and local elections. Although Midas storekeeper Arthur A. Primeaux ultimately lost his re-election bid to the Elko County Commission, he was the chosen candidate in Midas. M.J. Fitzgerald was tapped for Road Supervisor (Midas District); both were Democratic candidates. Reber went on to serve in the 1909 Nevada Assembly (which was decidedly Democratic) as the Assistant Chief Clerk.

Although Reber was no longer in town to exhort his readers to vote for Democrats, politics continued to interest the people of Midas. Information concerning the 1910 election is not readily available, but the results of the general elections for 1912 through 1922 provide some interesting insight into Midas voters.

For the 1912 election, 163 men registered, and 112 voted at the Midas precinct's official polling place, the schoolhouse. Many of their chosen candidates were affiliated with the Socialist Party. For President, the order of preference began with the Socialist candidate. Democrat Woodrow Wilson received the second highest number of votes and Theodore Roosevelt of the Progressive Party, the third. Votes for the incumbent, Republican William Taft, ran a very distant fourth.

In contrast, the majority of Nevada voters chose the Democrat. Roosevelt came in second. Nationally, of course, the winner was Wilson. Democrat Key Pittman was the choice of Midas voters for U.S. Senator from Nevada (he won). John E. Worden, the Socialist candidate, was their choice for U.S.

Congressman (he didn't).

The local offices of constable and justice of the peace were filled, and it appears that 1912 was the last time local voters chose the Road Supervisor (Midas District). Independent candidate F.A. Raup won the office.

Approximately 80 men voted in 1914, and most supported the Socialists. The continued popularity of the Socialist Party in Midas might be explained by the biting cynicism obvious in the following comments printed in the town's newspaper a few months before the election:

> The old time polititions [sic] will be with us in a few weeks, pating [sic] us on the back and leading us up to the bar, and with their hand always out, asking "give me," "give me," "give me" your vote for I need the job, and in a great many cases making promises that they never intend to fulfill.

Midas voters approved Socialist Party candidates A. Grant Miller and Martin J. Scanlon for U.S. Senator and Congressman, respectively. The Democrat candidates for Governor (Emmet D. Boyle) and Mine Inspector (Ed Ryan) were also Midas winners. However, of the four candidates tapped by Midas voters, only Boyle ultimately won the office he sought.

Interestingly, Midas voters did not support Andy J. Stinson's 1914 bid for State Mine Inspector despite (or, maybe, because of) his being the mine superintendent at the Eastern Star mine in 1912. The Elko *Daily Free Press* called Stinson "the popular manager of the Eastern Star," but this report came before the miners went on strike at the mine. The Republican Stinson received only seven votes--less then 10 percent--from the Midas electorate, but won the rest of the state.

Additionally, Midas voters--all men in the 1914 election--overwhelmingly supported extending the right to vote to women. On the statewide ballot question concerning women's suffrage, Midas men voted 69 to 8 in favor of the question. The rest of the state concurred, and Nevada women were granted franchise privileges six years before the passage of the suffrage amendment to the United States Constitution.

An interesting result of the passage of the national suffrage amendment concerned Midas seven years later. On March 24, 1927, the *Reno Evening Gazette* reported that the first woman to apply in Elko County for U.S. citizenship was Mrs. Edith Maria Lewis of Midas, explaining:

> Formerly it was not necessary for a woman to file papers but with the right to vote it was deemed just that they should accept this obligation. The law was passed in 1926 and now a woman must pass through the same procedure as a man to become a citizen.

The number of voters in 1916 rose 23 percent over the 1914 election, although part of the growth was due to women voting in a statewide election for the first time. The number remained high in 1918 (at 99 voters) and dropped only slightly during the next two elections (89 and 73 for 1920 and 1922, respectively).

Again, Midas voters chose some interesting candidates. Voters retained their interest in Democrats (a 1916 Midas news reporter remarked that "it is dollars to doughnuts that this town will go Democratic," which it did) as well as Socialists. Independent candidates also did well in Midas. Casting their ballots at the Gold Circle Club Building in 1916, Midas voters, unlike their fellow Nevadans, again chose Socialist candidates A. Grant Miller and Martin J. Scanlon for the offices of U.S. Senator and Congressman, respectively. In 1918 and 1920, the local electorate picked the Independent candidate for the U.S. Senate, Anne Martin. She was the first woman to run for this office in Nevada and the

Emmet D. Boyle (Courtesy of the Special Collections Department, University of Nevada, Reno

In the summer of 1914, Mabel Vernon and future U.S. Senatorial candidate Anne Martin (on right) campaigned throughout rural Nevada in favor of the right to vote for the women of the state. It is probable that they stopped in Midas as the voters there supported both the suffrage amendment and, later, Anne Martin. (Photo courtesy of the Special Collections Department, University of Nevada, Reno)

first to lose the election. By 1922, however, Midas ballots rarely favored third-party candidates. Despite this move away from unconventional candidates, Andy J. Stinson still could not get the support of Midas voters at the 1918 election, although he carried the rest of the state.

Another interesting candidate during this period was Ray Hanley, who ran for the Nevada Legislature in 1916. A Midas resident and a Democrat, Hanley was successful in his bid to represent Elko County in the 1917 Nevada Assembly. After his election, he married Irene Lyng, a former Midas schoolteacher, in Golconda and returned to Midas to live. The *Journal of the Assembly* for the 1917 session revealed that Hanley did not introduce any legislation specifically for the benefit of Midas. Hanley served one session and was the only Midas resident ever to represent Elko County in the Legislature. Although Noble H. Getchell moved to Midas during his state Senate term, he represented Lander, not Elko, County.

One of the most important political issues during this period was the prohibition of alcoholic beverages. Although the Prohibition Party's candidate for President received a whopping total of one vote from Midas voters in 1916, two years later they supported adding the Prohibition Amendment to the U.S. Constitution by a vote of 52 to 34.

As legally required, Prohibition was officially observed in Midas. For example, the contract for the sale of a Midas pool hall in 1927 specified that "no intoxicating liquors will be stored, sold or handled on said premises." However, former residents remembered the fairly common existence of stills and homemade chokecherry wine.

By 1926, Midas voters had changed their minds about Prohibition. Two questions on the ballot that year concerned the repeal of the amendment. As did the rest of the state, Midas voters sent a strong message that Prohibition was no longer wanted. But the unpopular law was not revoked until the passage of another amendment to the United States Constitution in 1933.

The number of Midas voters in general elections after 1920 ranged from a high of 153 in 1928 to a low of 17 in 1942. Democrats were still popular, but third party candidates no longer attracted the attention of Midas voters who began to vote more consistently with the rest of the state. Republicans began to win more votes, including Andy J. Stinson, who finally won in Midas in 1926. However, he was the only candidate on the ballot. When he was challenged in 1930 and 1934, Midas voters again picked his opponent.

During the last six years of the existence of the Midas precinct, the number of voters ranged from a high of 37 in 1944 to a low of 17 two years later. Twenty-six people voted in 1948 and 20, in 1950. The vast majority of the candidates who won in the Midas Precinct continued to be Democrats.

The election precinct was dissolved at some point after the 1950 election. By 1952, no more election results were reported from the Midas precinct, and Jim Murdock donated the town's ballot box to the Midas School Museum.

The Water System

Early in the development of Midas, the population was soon large enough to warrant the construction of a water system, one of the most important services for a Nevada mining town. In March of 1908, William Menken and C.W. Schein applied to the state engineer to appropriate water for reservoirs above Midas "by means of dam and pipelines and there be used for domestic purposes." A month later, the Midas correspondent to the Elko paper noted that "the water pipe has arrived and will be put in with all possible haste and the company will get their water from Big Spring up on the mountain, and no where in the world could better water be found." On April 22, 1908, the Gold Circle Water Company was incorporated.

In a 1984 interview, Gordon Warren recalled that his father, Charles S. Warren, bought the town's water system from the Salt Lake City company that had originally established the system. Apparently, the company acquired the necessary water rights from John G. Taylor. Warren eventually sold the system to Elmer Nunes; the exact date is unknown.

Many of the early published accounts concerning Midas asserted that water was plentiful. Warren and other residents, however, remembered that water was scarce, and the system did not supply running water to every house in town. Many of the homes relied on independent wells rather than the town system. Noreen Wilkerson Murdock recalled that washday was particularly difficult for the women of Midas as many of them had to transport water to their homes from the well in Arthur and Ellen Primeaux's yard. Even as late as 1957, only one home had an indoor bathroom; in most houses, water was piped just to the kitchen sink.

Although the old, wire-wrapped wooden pipes froze in the winter and went dry in the summer, the system has been operated continuously since 1908. John and Evelyn Sabin purchased the Gold Circle Water Company from Elmer Nunes in 1958 and operated it until Mr. Sabin's death in 1985. Four years later, Mrs. Sabin expressed her desire to release control of the system and deeded her interest to the Midas Water Cooperative.

This nonprofit organization was established to benefit all users of the system and to prevent any one individual from controlling it. On behalf of the community, Mert Mickelson presented a formal request for approval of the concept at an open meeting of the Public Service Commission (PSC) of Nevada late in 1989. On November 21, 1989, the PSC granted approval and transferred the Certificate of the Gold Circle Water Company to the cooperative.

Initial funding for repairs and maintenance of the system came from service fees and member assessments. In addition, the cooperative obtained federal and state money. In the waning days of the 1991 session, the Nevada Legislature passed Assembly Bill 572, sponsored by Assemblyman John Carpenter of Elko County, that appropriated $100,000 from the emergency fund for the supplemental city-county relief tax to Elko County for the drilling and equipping of a community well to serve the residents of Midas. Justification for this action was outlined in the bill and noted that the town's only source of water was a natural spring, which, at the time, was not producing "enough water to ensure the health, safely and welfare of the residents." The Midas cooperative also received a federal grant of $189,000.

By 1994, the cooperative had constructed most of the new system, which included a well, a 30,000 gal-

Midas in the 1930s. Notice the schoolhouse in the background and the houses' outdoor plumbing in the foreground. (Photo courtesy of Friends of Midas)

lon storage tank, new pipes, and fire hydrants. In the subsequent decade, the Cooperative received a $500,000 grant and a $130,000 loan from the Federal Government to improve the system with an additional 30,000 gallon storage tank and more fire hydrants. A new source for drinking water was identified, and Newmont Mining worked with the cooperative, a driller, and other local businesses and residents to develop the well. Currently, the cooperative has 51 members.

The Midas Water Cooperative holds its annual meetings during the town's celebration of the Fourth of July. All customers and the public are welcome. The officers elected for 2006-2007 were Mert Mickelson, President; Kris Mickelson, Secretary; Brian Elmore, Treasurer; Richard Bennett, Board Member; and Gene Casci, Board Member and Resident Manager.

It appears someone may have been left behind when Midas homes switched to indoor plumbing. (Photo courtesy of Friends of Midas)

Electricity, Telephone, and Internet

Unlike water service, the town did not always have a central power source, and residents often did not have electrical power for their homes and businesses. Food was cooked, homes were heated, and lights were lit by coal, kerosene, propane, wood, sagebrush, and other fuels. Electricity was simply unavailable.

During the last productive mining boom, electricity was sometimes provided to the town by the powerhouse, which was constructed with the Gold Circle Consolidated mill. After the mill was closed, the powerhouse (what was left of it after the 1941 fire) was dismantled. For over four decades, privately-owned generators, usually fueled by propane, were the only sources of electricity for homes and businesses in the town. Many people, however, did not own generators and relied on some kind of gas for light and heat. The *Free Press* noted in its special article about Midas in 1986 that it was the last town in Elko County without a central power system. The lack of electricity was unusual even in 1963 when the Winnemucca paper remarked that "Midas is one of the few towns, still inhabited, where no electricity is to be found."

On September 25, 1988, Sierra Pacific Power Company finally brought electrical power lines into Midas. The arrival of the company's services allowed residents to utilize appliances, such as washers and dryers, that generators could not handle and to install satellite dishes for television reception. Some residents considered that autumn day to be one of the most significant events during their Midas tenure!

Although telephone service has been available in Midas since 1908, it has not always included private lines. Through the 1960s, the town had only one telephone, which was a pay phone on Main Street. A few phones were added in the 1970s to the party-line system of Nevada Bell's Osgood Toll Region. During the next decade, the system was switched to private lines, and more phones were installed. In May 1993, Midas was part of a larger rural system sold by Nevada Bell to Oregon-Idaho Utilities. Currently, any residence or business may have a telephone of its own.

The internet came to Midas in 1998. The principal and teachers at McDermitt High School organized McDermitt-Humboldt Internet Provider (M-HIP) as a means to provide internet access to their students. The cost of subscribing to the satellite-based system exceeded the school's budget, so M-HIP sells dial-up access to anyone in northern Nevada with a computer and a telephone. For the residents of Midas, email is now a local phone call away.

All of the political functions of Midas—township, justice court, election precinct, and school district—were dissolved by 1972. But the mining district was still viable, and people were living and working in town. Existing utility services needed to be continued and new ones added. In fact, the technological advances of the late 20th century allowed utilities in Midas to be enhanced to a level of comfort not available to earlier residents. By 1994, the town enjoyed vastly improved water delivery, electrical power, television reception, telephone service, and fire protection. Four years later, internet access was added.

Although the town has not had any formal organization for over four decades, some residents have advocated the re-establishment of a political structure. The debate about setting up a Town Council to advise the Elko County Board of Commissioners began in 1997, and some have considered creating a general or special

Taken around 1970, this photo shows the poles that brought the one telephone line into town. (Photo courtesy of Nevada Historical Society)

improvement district. No formal action has been taken as of the time of this writing, however. In the meantime, in order to remain close to these constituents, the Elko County Commission has conducted a formal meeting in Midas annually since 1997 and continues to oversee community needs, such as sewage disposal and trash service. In 1997, the Federal Government closed the landfill that had served the town for decades. Trash service is now under the jurisdiction of Elko County, which provides a dumpster for local trash. The dumpster is emptied on a regular basis at a landfill in Humboldt County. Because many of the people who use the dumpster are not full-time residents of the town, collecting fees sufficient to cover the cost of the service has been a challenge for Elko County officials.

Two first graders and their teacher, probably Lela Halsell, in 1915.
(Photo courtesy of Lela Halsell Hempton, Friends of Midas Collection)

MIDAS SCHOOL DISTRICT

When Midas sought to open its first school in time for the 1908-1909 school year, residents discovered that the town would not be eligible for state school funds because Gold Circle had not been included in a state-mandated survey earlier in 1908. In the annual report to the State Superintendent of Public Instruction, the Deputy Superintendent for Elko County recommended that this law be changed as soon as possible because of the hardship it presented to new districts, such as Gold Circle's.

The Midas School District and Elko County cooperated to provide for the 30 or 40 school-aged children in the district, and school was opened by January 1, 1909. The Educational Directory of that date reported that H.W. Lane served as Clerk of the Midas School Board, and Miss Boyle was the teacher.

The 1909 report from Elko County noted that the schools of Elko, Wells, Carlin, Tuscarora, and Midas were considered "town schools"; the balance of the county contained "rural schools." Town teachers were paid an average of $89.23 per month; rural teachers, $60.66. Male teachers received $91.66; female, an average of $61.10. In 1910, the female teacher in Midas earned $75 monthly; in 1911, her salary was increased to $90. These salaries were considered low compared to teachers' salaries in the rest of the state, but comparable to miners' wages. Most mines paid around $4 per day.

Classes were conducted in the original school building, although some residents began to consider construction of another facility. The *Gold Circle Porcupine* asked in 1914:

Midas students and their teacher (possibly Thelma Johnson) pose in front of two different school buildings around 1917. It is assumed that one is the "Little School"; the other, the "Big School." Both buildings are long gone. (Photo courtesy of the Lyon Family, Friends of Midas Collection)

"How About The New School House?" Unfortunately, a printed answer to the question has not survived. Consequently, it is not clear if the query referred to a newly-built building or the discussion to construct one.

Two School Buildings

By 1916, the Midas School District enrolled 34 children and employed two teachers, indicating that the second school building had been completed. The first school building was located on a hill above the main part of town to the east and was probably built in 1908 or 1909. Five or six years later, a larger facility was built on a hill near the original school. For the next 10 years, the original building, called the "Little School," housed the lower grades (usually first through fifth). The new building, called the "Big School," contained the upper grades (sixth through eighth) and, occasionally, the first year of high school. Neither of these buildings remains standing.

The district was governed by an elected board of trustees. On May 31, 1916, the Elko paper printed the following report from Midas:

> Last week we were paid an official visit by Miss Kenemeyer, assistant deputy of public instruction. For the last six months this town has been a seething cauldron of discontent about the election of the school trustees for the ensuing terms. A mass meeting was held at Cody's hall with the result that F. Macy and R.T. Noble were appointed. Miss Kenemeyer made a most favorable impression and handled the situation with consummate skill and tact. At present all is quiet along the Potomoc.

Unfortunately, the details of the "discontent" were not provided nor have any results of school trustee elections been located, other than the names of the elected Clerks recorded in the *Educational Directory.*

School children walked to class from their homes in town and near the Elko Prince Mine, about 2 miles away. During the school years from 1919 through 1922, at least six children made that walk every day. The group was under the leadership of the oldest child, Violet Phillips (later Steninger), who was 11 years old when she and her family moved from Midas on the night the Elko Prince Mill burned in 1922..

The younger children were taught by Thelma Johnson (who married Ernest Clawson) in one one-room building. Grades 6, 7, and 8 were taught by Ethel Winger (who later married Miles Kenneth Doyle) in another. In the spring of 1918, the eighth grade graduating class consisted of best friends Marion Clawson and Bill Cody. Clawson long remembered the compliments he received for his recitation of Abraham Lincoln's Gettysburg Address at the graduation ceremony.

Neither the Clawson family nor the Cody family could afford to send their boys to the new boarding high school in Elko, so Miss Winger was their ninth grade teacher. In the same room with the sixth, seventh, and eighth graders, probably six or seven other children, the two boys were taught algebra, English, Latin, and ancient history. A graduate of the University of Nevada, Marion Clawson eventually served as the Director of the United States Bureau of Land Management in Washington, D.C. (1948-1953).

Marion's brother, Bill, recalled that the two teachers helped the town kids make a dirt tennis court. He explained that the women were "fresh out of Normal School at the University of Nevada, full of energy and really great teachers." During play time, the boys, including Bill's best friend John Cody, wore coveralls. At school, the boys, like Bill and his best friend John Cody, wore coveralls, and the girls wore dresses.

The *Nevada State Journal* recorded the final events of the 1919-1920 school year. Noting that school had ended on Friday, May 21, 1920, a dance at the "Midas auditorium" was held in honor of three teachers: Mrs. Leslie (Lelah) Noble and Mrs. Ray (Grace) Clawson of the Midas School and Miss Sadie Guill of the Meyers School. The paper remarked that "it was one of the most enjoyable dances ever held in the camp" and expressed mock sorrow about the children no longer having school to fill their days:

> For the next three months the boys and girls of the camp will have to undergo the inexpressible hardship of being unable to go to school and sit quietly for hours in joyous study when not reciting the well learned lesson. But the youngsters are brave and one would think they almost enjoyed the prospect of the long days when they will have nothing to do but play, and fish, and one thing or another. To compensate in a measure they were given a picnic last Saturday [May 22]. There were all kinds of good things to eat, including plenty of ice cream, and the children, as well as the grown folks present, enjoyed the outing immensely.

After the mill burned in 1922, the number of children attending the Midas School steadily declined. In 1926, the district requested just enough money to educate six children, the number then enrolled, during the next school year. But the revitalization of the Gold Circle Mining District soon caused the school population to increase.

From 1926 through 1929, two teachers were employed by the Midas School District. In 1928, 50 children enrolled at the Midas School, and Parent-Teacher Association meetings were common. Probably the most important activity in the school district during this period was the building of the new schoolhouse.

For the first years of its existence, the Midas School District was not known for its school accommodations. In the 1926 report to the State Superintendent of Public Instruction, the deputy superintendent for Elko County reported that "practically all the schools with two or more teachers have first-class buildings, Contact and Midas excepted."

A Two-Room School House

Midas children began the 1927 school year in the Big and Little Schools on the hill above town, but their elders determined that it would be the last year in those buildings. That fall, Midas voters approved a $12,000 bond measure to erect a new two-room schoolhouse. The 1927 district school board, consisting of Frank Macy, Chairman; Paul Ehlers, Trustee; and Charles Warren, Trustee and Clerk, published a notice of the intent to sell bonds to finance the new school building. Construction on the building began in November 1927 and was expected to be completed by the beginning of the school term in 1928. The debt on the building was retired in 1935.

In 1931, Midas students stand on the front porch of the new, modern Midas School. (Photo courtesy of the Lyon Family, Friends of Midas Collection)

In the next report, the deputy superintendent called the new school "modern in every respect" and noted that a "teacherage," to provide housing for the school's teacher, had been added in the fall of 1928. He proclaimed that "school conditions are now such as to induce good teachers to desire employment there with some degree of permanency."

In fact, one of the teachers who started at the new building in 1929, Mrs. Katherine M. Boyle, stayed through the 1931-32 school year, one of the longest tenures for a Midas teacher. Of the 21 known teachers from 1909 to 1928, only seven stayed more than one year. Of those, two stayed three years, one taught four years in total. After 1929, only four teachers taught in Midas more than one year. Throughout the history of the Midas School, only two teachers taught for four years, the longest period of employment. Thelma Johnson taught the younger grades from 1917 through 1919 and returned to teach under her married name (Mrs. Ray Clawson) in 1925. Mrs. Lois Pursell conducted classes from 1937 through 1941. In 1937, she was joined by Miss Viola Minneman, the primary grade teacher. That year was the last during which the Midas School District employed two teachers.

For most of these years, the school district prospered. In October 1936, the "Midas News Notes" announced the installation of playground equipment at the school. The columnist proclaimed that the "new swings at Midas grammar school help to make recess and noon more fun." Apparently, Mrs. Hilda Jacobson was instrumental in obtaining the equipment, which remains on the school property and is still enjoyed by kids of all ages. A year later, workers employed by the Federal Government's Works Project Administration (WPA) completed over $200 worth of work at the schoolhouse. Jack Murdock remembered that most of the work involved fencing the schoolyard.

The school's enrollment certainly reflected the fortunes of the town. Fifty students attended the Midas school for at least two years during this period; however, in 1933, only 16 children were enrolled. By September of 1938, 20 pupils were registered. This number appeared to remain relatively constant for the next few years.

Wayne Wilcox fondly remembered the graduation ceremony at the Town Hall for the eighth grade class of 1938, which consisted of him and Frances Macy (a Midas native who later married Joe Keller):

> I gave the valedictorian speech. My dog, Jerry, was sitting in the audience with my parents and when he saw me up on the stage, he came bounding up there and sat with me all during the ceremony. Everyone clapped when I finished. I think it was for old Jerry, though.

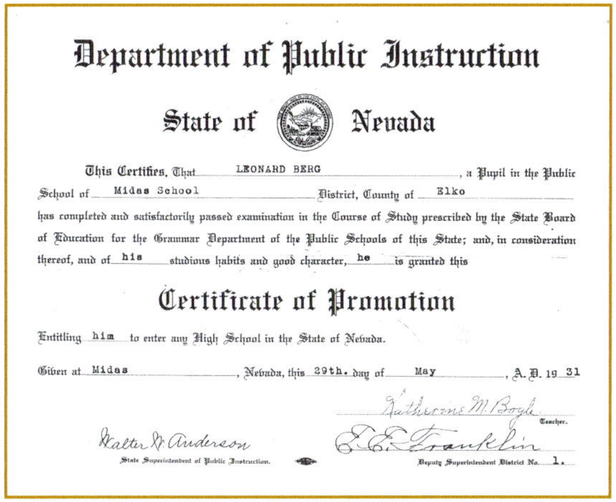

Department of Public Instruction

State of Nevada

This Certifies, That LEONARD BERG, a Pupil in the Public School of Midas School District, County of Elko has completed and satisfactorily passed examination in the Course of Study prescribed by the State Board of Education for the Grammar Department of the Public Schools of this State; and, in consideration thereof, and of his studious habits and good character, he is granted this

Certificate of Promotion

Entitling him to enter any High School in the State of Nevada.

Given at Midas, Nevada, this 29th. day of May, A.D. 19 31

Katherine M. Boyle, Teacher.

Walter W. Anderson, State Superintendent of Public Instruction.

E. E. Franklin, Deputy Superintendent District No. 1.

(Courtesy of Len Berg, Midas School Graduate, Friends of Midas Collection)

The Midas School District suffered with the mines when they began to decline after 1938. In March 1942, the "Midas News Notes" reported that numerous people were moving out of the district; only eight children were attending the town's school. Seven months later, it was observed that the number of schoolchildren had decreased to six, consisting of two groups of three students each from the Esmeralda Mine and nearby Squaw Valley Ranch. By the end of 1942, the school was closed because it could not enroll at least three students. It reopened, however, in 1943 and continued until at least 1951.

The *Educational Directory* stopped publishing the names of every teacher in Nevada in 1950. In 1955, the Nevada Legislature created county-wide school districts and eliminated the Midas School District along with hundreds of other tiny districts in the state. The Elko County School District did not maintain much historical information about the Midas school. Consequently, information about the operation of the school after 1950 is sketchy.

According to one of the first stories printed about the

These little girls appear to be ready for a spring pageant at the Midas School. (Photo courtesy of the Lyon Family, Friends of Midas Collection)

Top, left to right
Miss Purcell congratulates the eighth grade graduating class (Willie Wilcox and Frances Macy) in 1938. (Photo courtesy of F. Wayne Wilcox, Friends of Midas Collection)

These boys have been enjoying the snow in the schoolyard in the 1940s! The three in the back, from left to right, are Mickey Jensen, ? McKenna, and Richard Tracy; the two in front are unidentified and Kenny Tracy. (Photo courtesy of Gene Sweeney Gabica, Friends of Midas Collection)

Annie Jones stands in the west room of the two-room schoolhouse in 1966. (Photo courtesy of Joan Hall, Friends of Midas Collection)

Art Howe and teacher Laverl Schofield study in the Midas School in 1959. (Photo courtesy of Phil Gemmill, Friends of Midas Collection)

Howe and Schofield flank the stairs in 1959. (Photo courtesy of Phil Gemmill, Friends of Midas Collection)

ghost town of Midas, the school was closed in January 1952 "when the town no longer could supply the minimum of three pupils necessary to maintain it." The school reopened in 1954 with Dorothy Powell as the teacher. It is not known if the school was open or dormant for the next four years. In 1958, a family with nine children lived in the town, but they were home-schooled and did not attend the Midas facility, despite the assignment of Laverl Schofield as the teacher. Phil Gemmill described Schofield's arrival to oral historian Victoria Ford:

> Laverl, he graduated out of the university in Oklahoma, and they had these job openings in Nevada, and they had about four or five places. So he just pointed to Midas – no idea where it was. He didn't know it was twenty-nine miles of gravel road just to get there. And back in those days the road wasn't very good. [laughter] And so, I believe, he had a 1955 sedan, and he drove out there, couldn't believe that he was on the gravel road going nowhere. When he got there, there was no store, no place to live. He found the school. I believe he spent the first night in the school. But no restaurants. [laughter] I think Andy Jones ... No, they didn't even have the bar then. They were working there then. So my mom and dad took him in. But he just came there, and it was unbelievable. He didn't know *what* he was going to do. [laughter]

The school was open in the early 1960s primarily for the children who lived on nearby ranches. Harold Ridgeway, Assistant Superintendent for the Elko County School District, recalled that the school was open in 1963 when he arrived in the district and from 1967 to 1969 when he was the principal for the county's rural schools. The number of students rarely exceeded six or seven. According to Ridgeway, Mrs. Page was the last teacher in Midas.

The last child to complete all eight grades at the Midas School was Lester Sweeney, according to his mother and Midas Justice of the Peace (1942-1950) Maxine Sweeney.

The Midas School and the playground, featuring a witch's hat merry-go-round, as they appeared before restoration began in the late 1990s. (Photo courtesy of Friends of Midas)

Mrs. Meyers (standing in back) had a large class in the spring of 1960. The second girl (in the girls' row) is Marsha; Louise is at the end of the row. The first two boys at the front of the boys' row are Bill Blood and Ron Blood, respectively. The rest are unidentified. (Photo courtesy of Ron Blood, Friends of Midas Collection)

Midas Joint Venture erected this sign to commemorate the dedication of the newly-restored Midas School in 1998. (Photo courtesy of Friends of Midas)

In private ownership

On January 12, 1972, the Elko *Daily Free Press* reported that the school board had voted to close the Midas School permanently. Elko County School District Superintendent Robert Zander explained that the last students had left the area and recommended that the teacher be transferred to Carlin. The board requested an appraisal of the building, equipment, and lots. Five months later, the advertisement of the sale was authorized, noting that the minimum acceptable bid was $5,663. On July 11, 1972, the board accepted a bid of $6,000 from Dr. Tom Pitts of Reno, and the schoolhouse was sold. It remained in private ownership for the next two decades, functioning as a hunting lodge for the families in the Midas Gun and Country Club. Incorporated in Nevada in 1977, the privately-held club consisted of Pitts, Robert Britton, Gene Casci, William Higgs, Harry Huneycutt, Jim Megquier, Donald Richter, Thomas Trabert, and Donald Zundel.

In 1996, the group decided to sell the old building. The members of the Board of Directors for Friends of Midas discussed buying it, but determined that it would be too difficult to raise the necessary capital to buy and refurbish the aging structure. Ultimately, it was purchased by Midas Joint Venture, which originally planned to use the school building as its administrative offices. Knowing of the interest of Friends of Midas in the school, General Counsel Craig Haase promised to "treat the organization right."

The company did more than that. On March 19, 1997, Andre Douchane, Vice President of Operations, presented to the town residents the mine's plans for the building. The company proposed to renovate and restore the facility by patching and finishing the foundation; refinishing the wood floors and wood trim; installing new wiring, lighting, and a furnace; blowing new insulation into the walls; and painting it inside and out. The company also planned to clarify the encroachments onto property owned by the Unger and Vander Wende Families. Douchane explained that the company would use the building as an office, but not as long as had been originally planned, and would return the school to the community. A few months later, Douchane indicated that the company would establish a trust fund that would pay for the building's utilities for the foreseeable future. This generous allocation was continued by subsequent owners of the nearby mine.

A year later, on March 21, 1998, the Board of Directors of Friends of Midas formally issued, to the Elko County Commission and Midas Joint Venture, a resolution of acceptance for the school and schoolyard. The resolution stated that "Friends of Midas will manage and maintain the Midas Schoolhouse and associated property as a site for historic exhibits and for use as a community center as approved by the Friends of Midas Board of Directors."

During that intervening year, Midas Joint Venture invested $140,000 in the project and worked diligently to return the school house to its original splendor. The closet doors provide an example of the company's attention to detail. Both rooms had unique closet doors along one wall behind which the students hung their coats. The design allowed a draft to dry out the children's

Several Newmont employees made painting the schoolhouse their community project late in the summer of 2004. After they painted the lower part of the building, Newmont brought in a lift and a power sprayer to finish the job. (Photo courtesy of Friends of Midas)

The Midas School was beautifully restored in 1998. (Photo courtesy of Friends of Midas)

wet winter clothing. On the front side of these doors were chalkboards. In order to restore these doors, special hinges had to be made identical to the originals. The company located the one place in the country that could do the work, and the new hinges, made by hand, could not be distinguished from the old.

The mining company's staff used the building as an office until the administrative facilities were completed at the mine site, and on October 16, 1998, Midas Joint Venture formally transferred ownership to Friends of Midas. A gala celebration, planned by the two entities with assistance from the Nevada Mining Association, was attended by 120 people. The proclamation from Governor Bob Miller, read by Alan Coyner, Administrator of Nevada's Division of Minerals, designated the date as "Friends of Midas Day." Dale Erquiaga, former Director of Nevada's Department of Museums, Library and Archives, served as Master of Ceremonies, and speeches were made by Russell Fields, President of the Nevada Mining Association; State Senator Dean Rhoads; Midas Joint Venture executives Douchane and Haase; and Friends of Midas President Dan Bennett. Twenty-two former students attended, including Edna Wilkerson Timmons who was the first to graduate from the eighth grade in that building (1928) and Lester Sweeney. By the time of the event, Midas Joint Venture had installed display cabinets on the museum side of the building, and the exhibits were in place. The formal activities at the Midas School were capped by a public lunch at the George DeLong garage and an evening of dancing in the school house.

Immediately, the building became the center of numerous activities for the residents and visitors to Midas and the surrounding area. Many people enjoyed perusing the Midas School Museum, which continued to grow with additional exhibits. The facility and its beautifully landscaped grounds hosted many events, such as the annual Fourth of July celebrations and Elko County Commission meetings, and provided a base for important activities such as fire-fighting crews.

During the restoration work, no major structural changes were made. Fire codes and insurance coverage demanded the shingles be changed from wood to a flame retardant material. However, the color of the replacement shingles was the same as the original wood shingles. On the interior, care was taken to match stains and paints with those used originally. The bell had disappeared long ago, but was replaced with a #20 cast-iron bell, and the belfry was rebuilt to exactly the same dimensions of the original. Included in the restoration was landscaping and a lawn watered by an automatic sprinkler system. A thermostat-controlled, forced-air furnace occupied one of the coal bin areas. The original wood and coal burning stoves had been removed many years earlier.

In 2004, Sierra Pacific Power Company granted funds to replace the windows with thermal pane glass, and Newmont Mining provided the equipment and labor to repaint the entire building. Friends of Midas bought the paint. Lawn maintenance was assisted by equipment donations from Gus and Lori Wegren and Tony and Paula Scheuller. Joan Bennett was often found wielding the weed-eater around the edges of the yard.

The Midas School building consisted of two large rooms. Originally intended to house the lower grades on one side and the upper grades on the other, the east room functioned as a community center, providing a place for formal meetings and morning aerobics classes.

The west room housed a museum of Midas memorabilia. Over time, items that had once been in the school found their way home, such as textbooks donated by Keith Paskett and a globe donated by Jennifer St. Germain. In

The swings and other playground equipment have been popular with kids of all ages. (Photo courtesy of Friends of Midas)

Friends of Midas developed a small museum in the west room. The piano in this picture had been in the Midas Town Hall for many years and in the Midas Bar for many years after that before being moved to the museum in 1998. (Photo courtesy of Friends of Midas)

The exhibits in the museum included many items that had been used or found in the Midas area over the years. (Photo courtesy of Friends of Midas)

addition to historic school items, the museum displayed mining equipment, the Midas Precinct ballot box, the piano from the old Town Hall, an old Nevada flag loaned by the Northeastern Nevada Museum, and many other pieces donated by Midas enthusiasts. Both rooms contained numerous historic photographs, including Kirby Amthor's extensive collection from the 1970s and 1980s.

In 2004, the Elko County Chapter of the Nevada Archaeological Association was commissioned by Friends of Midas, with seed money provided by Keith and Sheri Thompson and Tony and Paula Scheuller, to construct an exhibit with a diorama depicting the life of prehistoric people in the area. Installed in the fall of 2005, the diorama illustrated hunting, stone quarrying, food gathering, and other activities that may have occupied the lives of early inhabitants around 500 years ago. The exhibit also included a display of various tools and manufacturing methods.

On August 12, 2004, State Historic Preservation Officer Ron James announced the listing of the building on the National Register of Historic Places. Earlier in the year, it had been listed on Nevada's State Register of Historic Places.

Then the inconceivable happened. Late on November 6, 2005, a fire of indeterminate origin started somewhere in or around the Midas School. By the time the flames were spotted after midnight, the fire was out of control. Terry Bates, who had seen the fire from his house, called Gene and Barbara Casci. Mr. Casci obtained the truck from the fire house and sped to the school house. Mrs. Casci telephoned others for help, but the hastily assembled firefighting team was unable to save the building. They were, however, able to keep the flames from spreading, especially to the propane tanks nearby. Instrumental in the effort were Bates, the Cascis, Mike Conley, George DeLong, Midas Volunteer Fire Chief Dwayne and Molly Jackson, Ken and Bert Klamm (who had also seen the fire from their house), Jerry and Garla Singleton, Mike White, Jim White, and Steve and LaDonna Zimmerman. Investigators from the Nevada Division of Forestry arrived on the scene within hours.

Saved were the outhouses (believed to have been built by the Works Project Administration during the Great Depression) and the much-loved playground equipment. The witch's hat merry-go-round is one of the last of its kind in the United States.

Lost with the building were the historic items and pictures on display in the museum. All that were found in the ashes were a few crucibles from the mining display and all but one of the metal letters from the sign that hung over the door.

The destruction of this building is a blow to the many people who attended school there, who enjoyed the playground, who visited the museum, and who carefully cared for this wonderful facility. It had become a symbol of the town and a point of pride for everyone who works hard to preserve Nevada's quickly disappearing historical buildings.

At the time of this writing, plans to rebuild on the site are underway.

(Photo courtesy of Friends of Midas)

A surreal sight. (Photo courtesy of Bert Klamm, Friends of Midas Collection)

MIDAS SCHOOL DISTRICT, 1909-1951
SCHOOL BOARD CLERKS, PRINCIPALS, AND TEACHERS

Year	Clerk	Principal	Salary	Teacher	Salary	Total # of Students
January						
1909	Mr. H.W. Lane			Miss Boyle	NR	
1910	Mr. H.W. Lane			Miss Irene Lyng	$75	
1911	Mrs. Katherine Boyle			Miss Irene Lyng	$90	
1912	Mrs. Katherine Boyle			Miss Anna M. Mack	$90	
1913	Mrs. Katherine Boyle			Miss Katherine Lyng	NR	
1914	NR			NR		
1915	NR			NR		
1916	Mr. Richard T. Noble	Miss Irene Lyng	$1,000	Miss Lela Halsell	$80	29
1917	Mr. Richard T. Noble	Miss Carna H.Damm	$100	Miss Thelma C. Johnson	$85	24
1918	Mr. Richard T. Noble	Miss Ethel Winger	$100	Miss Thelma C. Johnson	$85	21
1919	Mr. Richard T. Noble	Miss Ethel Winger	$100	Miss Thelma C. Johnson	$100	23
Nov/Dec						
1919	Mr. Richard T. Noble	Miss Rose L. Tolley	$990	Mrs. Lelah B. Noble	$900	25
1920	Mrs. Anna Hargrove	Mrs. Grace E. Clawson	$1,125	Mrs. Lelah B. Noble	$1,035	30
1921	Mrs. Genevieve A. Morey			Mr. W.B. Robinson Mrs. Elizabeth Robinson	$1,350 $1,260	
1922	NR			Mrs. Francelia Proebstel Miss Greta Wells	$1,350 $1,125	
1923	Mr. Charles S. Warren	Mrs. Francelia Proebstel	$1,350	Miss Greta Wells	$1,125	
1924	Mr. Charles S. Warren	Mrs. Francelia Proebstel	$1,500	Mrs. Ina E. Angus	$1,250	
1925	NR			Mrs. Thelma J. Clawson	$1,350	6
1926	Mr. Charles S. Warren	Mrs. Florence L. Peacocke	$1,350	Miss Ada Sharkey	$1,215	28
1927	Mr. Charles S. Warren	Miss Margaret Watson	$1,350	Miss Alice Williams	$1,215	33
1928	Dr. Daniel H. Pettingell	Miss Margaret Watson	$1,500	Miss Vesta M. Ferry	$1,350	31
1929	Mrs. Ellen Primeaux	Miss Jessie I. Diamond	$1,500	Mrs. Katherine M. Boyle	$1,350	18
1930	Mr. Frank M. Macy			Mrs. Katherine M. Boyle	$1,650	12
1931	Mr. Frank M. Macy			Mrs. Katherine M. Boyle	$1,250	16
1932	Mrs. Beth Finkle			Miss Florence Webber	$1,125	24
1933	Mrs. Beth Finkle			Miss Adele Myers	$1,080	17

Year	Clerk	Principal	Salary	Teacher	Salary	Total # of Students
1934	Mrs. Hilda Jacobson			Miss Bess Hoover	$1,125	30
1935	Mrs. Hilda Jacobson			Miss Ina G. Sharp	$1,125	20
1936	Mrs. Hilda Jacobson	Mrs. Juanita Brown	$1,170	Mrs. Irma Thornby	$1,125	20
1937	Mrs. Hilda Jacobson	Mrs. Lois W. Pursell	$1,170	Miss Viola Minneman	$1,080	31
1938	Mrs. Hilda Jacobson			Mrs. Lois W. Pursell	$1,350	21
1939	Mrs. Ellen Primeaux			Mrs. Lois W. Pursell	$1,350	19
1940	Mrs. Ellen Primeaux			Mrs. Lois W. Pursell	$1,350	18
1941	Mrs. Ellen Primeaux			Mrs. Isabella S. Happal	$1,350	21
1942	Mrs. Ellen Primeaux			Mrs. Ann Fettic		6
1943	Mrs. John Churchfield			Mrs. Florence Myers	$150	6
1944	Mrs. Paula Boyles			Mrs. Mary Lee Duncan	$1,575	9
1945	Mr. Paul Sweeney			Mrs. Florence Myers	$1,575	6
1946	Mrs. Noreen Murdock			Mrs. Corrine Hickey	$1,575	6
1947	Miss Ardis Wilkerson			Mr. Henry T. Imes	$2,400	8
1948	Mrs. Noreen Murdock			Mr. Henry T. Imes	$2,400	10
1949	Mrs. Noreen Murdock			Mr. Henry T. Imes	$2,400	9
1950	Mrs. Clyde Green			Mr. Dale D. Wolfe	$2,400	8
1951	NR			Mrs. Sadie Andrae	$2,400	8

NOTES: Data sources: *Educational Directory*, Elko County School District
The principal was also employed as a teacher, usually for the upper grades (6-8).
The teachers instructed grades K-8. In 1916, 1917, 1918, and 1937, ninth grade was also taught.
Amounts at $150 or less indicate monthly salaries; larger numbers indicate annual salaries.
Although the school district had a teacherage, only the 1939 report indicated that the teacher received salary plus living quarters.

NR - No report.

MIDAS SCHOOL DISTRICT, 1951-1972
STATUS OF SCHOOL
TEACHERS' NAMES

Year (Sept.)	**Status of School**	**Teacher**
1952, 1953	Dormant	
1954	Open	Mrs. Dorothy Powell
1955, 1956	Unknown	
1957	Dormant	
1958	Open	Mr. Laverl Schofield
1959	Dormant	
1960	Open	Miss Willet Myers
1962	Unknown	
1963	Open	Unknown
1964	Open	Mrs. Janice Benninghove
1965, 1966	Unknown	
1967, 1968, 1969	Open	Unknown
1970	Unknown	
1971	Open	Mrs. Lola Page
1972	Midas School sold to a private party	

Data source: Elko *Daily Free Press*, Elko County School District

A blanket of snow silenced the playground at the Midas School.
(Photo courtesy of Friends of Midas)

By 1910, the Wilkerson brothers were no longer in partnership with Arthur Primeaux in this general store located on Main Street, but they may have remained partners in the delivery part of the business as indicated by the wagon. The windows advertise "Drugs"; the sign over the door announces the presence of a Notary Public. Note the add-on to the building. This was the original post office; it was accessed through the store. On the store's left is the Capitol Saloon. (Photo courtesy of the Northeastern Nevada Museum)

BUSINESS LIFE IN MIDAS

The first blush of mining activity and the growing population created a thriving business community early in Midas. In April 1908, the Elko newspaper related that heavy freight wagons loaded with supplies and pulled by eight- and ten-horse teams were arriving in the new camp at the rate of 10 to 15 per day. That same month, the *Battle Mountain Herald and Central Nevadan* provided a graphic description of the amount of goods being sent to Midas:

> The teams that left Battle Mountain loaded with lumber for the Circle looked like a circus parade. There were over 20 six- and eight-horse teams all in a line moving across the flat at one time. There is about one-half a million feet of lumber here now for the Gold Circle Mercantile Company that will be moved into the camp as fast as teams can haul it. There are over 150 head of horses on the job now. The sidetracks of the Southern Pacific are pretty well filled up with carloads of freight for this camp. There are three cars of beer, two of grain, two of flour, 16 of lumber and one of groceries.

The interior of the Primeaux Store around 1911. (Photo courtesy of Edna Wilkerson Timmons, Friends of Midas Collection)

Obviously, Midas residents had their priorities!

These provisions built homes and supplied businesses in Midas. In April 1908, the Battle Mountain paper noted that the business community of Midas consisted of:

> Nine lodginghouses [one had 30 apartments], 11 saloons, five grocery and general stores, two lumber yards, two assay offices, two real estate offices, one bakery, two news and confectionery stores, two butcher shops, two feed stables and numerous other enterprises.

Advertisements in the local newspapers provided the names and locations of some of these establishments. Newly arrived

Ed Wilkerson on Main Street early in Midas's development. Behind him are the Primeaux-Wilkerson Company store (offering hardware, drugs, and groceries) and the Capitol Saloon. (Photo courtesy of the Northeastern Nevada Museum)

prospectors could obtain needed supplies at one of the general merchandise stores, such as Maney Brothers and Company, on the corner of Second and Main Streets; Primeaux Wilkerson Company, near Fourth and Main; Bridges' Grocery Store on North Main Street, or the Gold Circle Meat Market on Main. They could buy drinks and cigars and do some gambling at a number of saloons, including The Crystal, The Capitol Saloon, The Eureka Saloon, The Northern, Gold Dust Saloon, The Club Saloon, Elko Saloon, and Midas Saloon. One of the other bars, the Gold Circle Club, advertised that it was "The Largest and Best Equipped Resort in Midas" and offered nightly vaudeville performances.

A good meal could be obtained for 50 cents at places like the Silver Dollar Restaurant, the Canyon Café, Charters' Restaurant, or the T-Bone Restaurant. The McDonald Restaurant, located at the corner of Canyon and Main, advertised in the local paper: "Meals Served at Regular Hours. Home Cooking a Specialty. Prices Moderate." Clothes could be cleaned at the OK Laundry or the Gold Circle Laundry on South Main Street. Bodies could be washed at Burge & Heppner. A sweet tooth could be satisfied at the Confectionery on Main Street, and connections with the outside world could be maintained through the Postoffice News Stand or John Mortimer, "Public Accountant, Stenographer and Typewriter, Telephone Office,"

Weary miners and travelers could find a room (which was, most likely, no more than a bed separated from the next by a canvas partition) at the Hotel Monte Carlo on the corner of Main and Second Streets; the Hotel Nevada, a two-story frame building on North Main Street; the Gold Circle Hotel, which could accommodate 100 guests; the European Hotel, which advertised that it was "fitted with iron beds"; the North End Lodging House on North Canyon Street; the Canyon Hotel on North Main, which had "Clean, New Beds" for 50 cents and up per night; or the Grand Hotel, "Cleanliness Our Motto," where

The Maney Brothers Store was one of the first businesses in Midas. It is not known if these are, indeed, the Maney Brothers sitting on the stoop of their enterprise around 1908. Although it looks like a brick building, it was actually covered with a material designed to resemble brick. This building, located across Main Street from the Primeaux Store, was acquired by Charles Warren who continued its operation as a store for many years. The building burned in 1964. (Photo courtesy of Edna Wilkerson Timmons, Friends of Midas Collection)

cots were 50 cents and beds one dollar per night.

Other enterprises found in the district included an attorney (on the corner of Fourth and Main), three notaries public, one contractor and builder, a dentist, three assay offices, and livery and feed stables. Obviously, most of these businesses advertised in the local newspapers. Some Midas businesses also advertised in the Battle Mountain paper, including Blossom & Blossom (butchers), Liddell & Liddell (mining engineers), and The Midas saloon.

Many other businesses were enthusiastically planned but never built. One of these projects included a bank. The Battle Mountain paper reported in December 1907 that Elko businessmen were planning to open a bank in Midas. They had even gone so far as to purchase lots from the Gold Circle Townsite Company for the proposed 25- by 75-foot, two-story, stone building. Ultimately, the bank never materialized.

Most of the men, obviously, were engaged in

Charles Warren arrived in the Gold Circle Mining District early and stayed for many years. (Photo courtesy of Desda Warren Wood, Friends of Midas Collection)

Arthur and Ellen Primeaux in front of their store, probably in the early 1930s. (Photo courtesy of Friends of Midas)

mining and mining-related jobs. The 1910 report from Nevada's Inspector of Mines includes specific employment information for four of the mines in the district. The Sleeping Beauty employed eight men; the Rex, ten; the Gold Circle Crown, six; and the Elko Prince, two. The report noted further that miners were paid $4 per day and engineers, $5. It also related that all of the men employed at the Beauty and the Rex were Americans. No such comment was included for the employees of the other two mines. In 1920, 78 percent of the men were employed in such activities, but wages had not increased. The 1930 census information reveals the beginning of a decrease in the percentage of Midas residents engaged in mining. Never again did most of Midas's residents work in the mines.

After 1908, Midas businesses did not advertise in any of the regional publications. Consequently, the census reports are the best resources for many of the enterprises found in the town. The 1920 census recorded two grocery stores, three boardinghouses, a cobble shop, one general mercantile, a barbershop, and a confectionery store. In addition, the existence of a saloon was noted, despite the fact that Prohibition was still in effect. The 1930 census listed a grocery store (operated by Charles Warren), a barber shop (operated by Frank Macy), a general store (operated by Arthur Primeaux), a soft drink establishment (operated by Dick Vanatta), a shoe and harness shop (operated by Texas Wilson), and a mechanic shop (operated by D.L. Young).

The Primeaux store, the only business mentioned in all three census records, became the longest-running operation in Midas. In 1908, Arthur A. Primeaux, then an Elko County Commissioner, teamed with fellow Tuscarorans Ed and Louis Wilkerson to establish a general merchandise and grocery store. By 1910, Primeaux was the sole proprietor of the store, which he and his wife, Ellen, operated for over 30 years. During the years that their son Loring was in school in Reno (early 1930s), Victor Jacobson ran the store.

The number and names of local businesses in the town during the 1920s and 1930s are difficult to determine. This was a busy and prosperous time in Midas, but no local newspaper was published, and none of the establishments advertised in the *Battle Mountain Scout* or any other regional paper. Former residents recalled restaurants, bars (when they were legal and sometimes when they were not), pool halls, an ice cream parlor, a barber shop, automobile repair shops, and boardinghouses. Some of these operations provided room and board; some, only meals, which would cost a mining man about $37 a month. Two general stores operated almost directly across Main Street from each other: Primeaux's store and the Warren Mercantile. Gordon Warren also remembered two ice houses, which were supplied by a pond at the mouth of the canyon. He recalled men going to the pond in the wintertime to cut the ice, which was then stored in the houses for use during the summer.

Although it did not mention many businesses, and few by name, the Midas column in the Battle Mountain paper was useful in providing information about some of the enterprises operating during this period. In 1926, the column noted that the hotel on Main Street had been sold. Two pool halls were mentioned in 1928. In 1933, the columnist reported that "Mr. and Mrs. P. Murphy have opened the Corner pool hall and are serving candy, soft drinks and ice cream." A 1934 article related that Mrs. Barney (Beth) Finkle closed her restaurant; later that year, the columnist mentioned that the restaurant had a new owner, but did not indicate if these restaurants were two different establishments. Also in 1934, the "Midas News Notes" informed readers that J. Wilson had opened a butcher shop "in the tin house" and was selling vegetables. This establishment closed in 1936. The first mention of a bar is in a 1935 item, which observed that the "Murphy saloon" was being operated by the Rossi brothers. The pool hall was scheduled to reopen, according to the reporter in

Wes Allen had the only bar in town, the Acme Bar, in the mid-1950s. (Photo courtesy of Friends of Midas)

Wes Allen, in overalls, stands in front of his bar with Dick Lego on his left (in the white, long-sleeved shirt). The others are unknown. At the time this picture was taken, Lego was married to Lois Landis. (Photo courtesy of Friends of Midas)

In 1949, Pat Sweeney and Ruby Warren pose with the town's mascot, which, rumor has it, enjoyed a good beer! (Photo courtesy of Friends of Midas)

By the late 1960s, the Midas Bar was the only business operating in town. (Photo courtesy of Friends of Midas)

October 1936. Other business references included the operation of a boarding house by Mae McKenna and a restaurant at the Esmeralda Mine in 1939.

After 1942, the only business open in town has been a bar to serve the occasional miner (when a mine was open and working), the more frequent cowboys, and the increasing numbers of tourists. During the 1940s, Lou Wilkerson and the Warrens, Gordon and Ruby, operated competing bars. But it was Wilkerson's bar that attracted the big-name entertainment. Former residents recalled that Bing Crosby, who bought an Elko County ranch in 1943, visited Midas on occasion and sang duets with Wilkerson.

In 1952, a ghost town chronicler commented about a bar in Midas, but did not provide the name or exact location of the business. Wes Allen operated the Acme Bar in the mid-1950s; it stood on the northwest corner of Main and Fourth. Another bar was built by Bill Vest and his son in the spring of 1958 and stood, for over 40 years, on the east side of Main Street, a few buildings north of Fourth Street, guarded by a sign warning "no minors aloud" [sic]. When the building was demolished, the sign was donated to Friends of Midas. Apparently, this operation was open for only a few years; for a brief time in the early 1960s, it was run by Richard Robertson.

The bar that most recent visitors recognize as the Midas Bar did not open until around 1958. Evelyn Sabin recalled that Andy Jones, after arriving in town in 1957, acquired the Primeaux store building, which he tore down and rebuilt just north of its original location. He moved a barback from Golconda into the building and opened for business as a combination bar/café/gas station. After the death of Andy's wife, Annie, in 1972, the business was bought by Bob and Ida Unger of Winnemucca. Two other owners briefly followed the Ungers in the mid-1970s, then Kirby Amthor purchased the Midas Bar in December 1977. For several years, Kirby's Midas Bar was the primary location for square dances, afternoon card games, potluck parties, and gatherings of hunters, snowmobilers, and, of course, thirsty Midas residents and visitors. Recently, Amthor donated to Friends of Midas over 1,300 photographs taken during his ownership of this popular establishment.

Amthor sold the Midas Bar to Les and Bev Matson in October 1984. They renamed it the Midas Bar and Pack Station and reopened the café part of the operation for the first time since Jones's ownership. Known as the Midas Saloon and Dinner House, Les received statewide press coverage for his unparalleled meals. In August 2006, the Matsons sold the Saloon and Dinner House to George and Terri Watford who, along with their son Matt, are quickly establishing a reputation for providing a comfortable dining experience for Midas residents and visitors.

True to the history of Midas, the town that has never been too small for at least one bar, a competing saloon opened in 1983. Called The Kountry Store, this bar was operated by Sam and Edythe Seals and was the site for community parties for many years. One of the most popular parties was the annual Halloween/ Nevada Day bash. It was often full on New Year's Eve, too. By 1997, this bar had been purchased by Don and Laverne Mellen and renamed the Midas Gold Circle Bar. The Mellens closed the business on July 1, 2005, and moved from the area. It is currently for sale.

Andy Jones rebuilt and operated the "new" Midas Bar from 1969 to 1972. (Photo courtesy of Friends of Midas)

Dan Bennett and the Midas Bar as they appeared in the 1970s. This telephone booth was the only phone in town for many years. (Photo courtesy of Friends of Midas)

This bar, built by Bill Vest, operated sporadically in the late 1950s and early 1960s. It was long abandoned and demolished in 2005. (Photo upper left, courtesy of Friends of Midas; upper right, courtesy of Roy Powers)

In 1986, the Kountry Store let in anyone who wanted to party on Halloween, provided that he, she, or it were in costume. This horse entered the bar dressed as Paul Bunyan's Big Blue. (Photo courtesy of Friends of Midas)

For many years, the abandoned bar warned that no one under 21 was "aloud." The sign was donated to Friends of Midas before the building came down. (Photo courtesy of Friends of Midas)

Women in Business

From the town's first year, women also operated businesses and worked for wages in Midas. According to 1908 advertisements in *The Gold Circle News*, at least five women were engaged in commercial activities. Mrs. Budgett operated the Hotel Nevada; Mrs. McKenna, the North End Lodging House. Mrs. McDonald was the "proprietress" of the McDonald Restaurant as was Mrs. C.A. Harrington of the Confectionery. Mrs. E.C. Hills ran a lodging house that offered "nice clean beds at reasonable prices."

According to census information, though, most women in Midas did not have jobs. Homemaker and mother were not considered occupations and were not listed. In 1910, only 11 women (less than 29 percent of the female population) stated a vocation. Their titles consisted of dressmaker, waitress, laundress, merchant, hotel operator, dentist, assistant postmistress, and cook. Among the working women of 1920, four operated boardinghouses; two were teachers. Other wage-earning women included a trained nurse and the postmistress, Miriam Stenson. Only two of the employed women were married. Stenson, a widow, was also listed in the 1910 census as running a restaurant. In an interview in 1952 (still living in Midas but with the last name Purdy), she said that she had been a young widow with two small children to raise when she arrived in 1907 to run a boardinghouse for miners.

In 1930, employed women consisted of one postmistress (Isabel Lyon), one public school janitor (Marie Ehlers), two public school teachers (Jessie Diamond and Katherine Boyle), and two who ran boarding houses. These two women were the only employed women who were married and living with their husbands; interestingly, no boarders were listed in these two households. However, Wayne Wilcox recalled that, in 1936, "the single miners ate at Mrs. Eckman's boardinghouse and lived in the little shacks all over Midas." Another boardinghouse, the one at the Esmeralda Mine, was operated by Mae McKenna in the late 1930s. She may be the same Mrs. McKenna who advertised the North End Lodging House in 1908. At the Winnemucca funeral of Mrs. Mary McKenna on June 27, 1949, all of the pallbearers were current and former Midas residents.

Despite the shrinking of the Midas business district over the decades, women continue to operate successful businesses. In the 1980s, Edythe Seals was the colorful bartender at The Kountry Store, and Bev Matson was often found in the restaurant at the Midas Saloon and Dinner House.

Women's involvement in mining has changed over the years, too. Not one woman was listed in the census details as being employed in mining, and no former resident mentioned female miners. Currently, however, 13 women (about six percent of the total) are employed at the Midas Mine. Their jobs are not just in the office, either: five operate underground equipment, one is a mine geologist, and one, Jennifer McMullan, is the Chief Mine Engineer. A Newmont manager explained the increasing number of women employed in mines:

> Women often become better scooptram and truck operators than men because of attention to detail on pre-op equipment checks, less damage to equipment, and more consistent performance/output.

The owner of the Midas Café, also known at the Lewis Café, treated the town's children to dinner and gift bags of candy and oranges for Christmas. This picture was taken after one of those dinners in the early 1920s. Against the café's wall, from left to right, are Theodore Caswick, Charlie Thompson, Jim Castor (smoking), unidentified girl, Don Hargrove, Gordon Warren (with hands up), Laura Wilkerson, unidentified boy and girl. In the next row, left to right, are Parkinson, Jenny Lewis, Desda Warren, Lucille Parkinson, Leroy Noble, Byron Wilkerson (only the side of his head is visible), unidentified boy. In the front row, from left to right, are Eleanor Coleman, Edna Wilkerson, Noreen Wilkerson, Mike Paisly with his arms around Loriene and Irene Macy, Edith Ehlers (with bear), Ernest Lewis, and Willy Lewis. (Photo courtesy of Edna Wilkerson Timmons, Friends of Midas Collection)

Edith Lewis ran the Midas Cafe from 1920 to 1928. After Mrs. Lewis left Midas in 1928, Edna Wilkerson Timmons helped her mother run this restaurant, which was located on the northeast corner of Fourth and Main. It burned in 1933, leaving the safe that still stands on the corner. Taken in 1930, this picture shows Edith's daughter, Jenny Lewis Vest, and her daughter, Mary Lou, next to the restaurant. (Photo courtesy of Mary Lou Vest Smith, Friends of Midas Collection)

The Red-Light Districts

Other well-known occupations for women, though also not one listed on any census, were prostitute and madam. It is reasonable to assume that one or more of the women recorded in each census were employed in the town's red-light district, one of the first businesses established in any mining camp. The Midas district, which rarely consisted of more than one building, moved several times. Its first move took it away from the main part of town, but that location ended up being too close to the planned site of the new schoolhouse in 1927. The women then moved to the other side of the town's center and up the canyon. The district continued to relocate farther up the canyon until it was the last house on the main road.

Although prostitutes, and the madams for whom they worked, came and went with great frequency, the red-light district appears to have been occupied by at least one woman from the beginning of the town through the World War II period. Four of the names that have endured are Ol' Dutch Annie, Silvertip, Irene True, and Lois Landis.

Ol' Dutch Annie was one of the first prostitutes in the area. In a 1984 interview with Sue Ann Stevens, Nona Trembath recalled that her husband, Benjamin, required the sheriff's assistance to compel Annie to answer the 1910 census questions. The data compiled from the census included an entry for Annie Singleman who was 35, divorced, and an immigrant from Germany in 1889. She lived alone and listed her occupation as "dressmaker." Perhaps she had an additional occupation.

Silvertip was remembered as a quiet, refined lady who was well-regarded by the townspeople. Several

The Rock House was the last red-light district still standing long after the women had left. By 1979, when this picture was taken, it had been in private ownership for almost two decades. Like other Midas buildings, fire destroyed the Rock House. (Photo courtesy of Friends of Midas)

Silvertip on the front step of the Rock House. (Photo courtesy of the Lyon Family, Friends of Midas Collection)

remembered her first name as Marion, but not all agreed on her surname nor could explain her unusual moniker. One former resident recalled that the townspeople helped Silvertip conceal her occupation from a visiting relative. She was one of the madams of the Midas red-light district during the late 1920s and into the 1930s, apparently operating out of the Rock House, which was the last house up the canyon until it burned around 1980.

According to former Midas residents, Irene was a madam in the 1920s until she married Curtis True in 1927 and moved to his home in Winnemucca. Remembered as a tall, strong woman with strawberry blonde hair and stunning diamond earrings, True drove the truck between Winnemucca and Midas for her husband's freight and mail delivery service. She habitually carried a shotgun for protection and often stopped along the way to do some prospecting, which was more than just a hobby. In 1938, the mining section of the *Reno Evening Gazette* reported that True, a "local woman mining promoter," had recently located promising claims in Humboldt County's Leonard Creek Mining District. Her obituary noted, later, that she had been "active in business and mining ventures." Curtis True died in Winnemucca in 1949, and Alma Irene True died in Reno in 1966. Both were buried in the True family plot at Reno's Hillside Cemetery.

After the Trues were married, Lois Landis took over Irene's house, which stood on the property currently owned by Bob and Ida Unger. Lois may also have had a different surname when she worked as a madam in Midas as former residents remembered her marriage to Jack Landis.

Determining the number of women who worked in the red-light houses is difficult. Former residents recalled that at least two or three women occupied the district when the mines were booming; sometimes, only one prostitute was in the area during the difficult economic times. Jack Murdock asserted that the town's madams were Silvertip and Lois Landis when he arrived in 1933. Wayne Wilcox remembered that 1936 Midas had "two grocery stores, Warren's and Primeaux's, Hungry Joe's Saloon and five prostitutes for the single miners (and anyone else)."

The residents' stories indicated that the existence of the houses was tolerated by the rest of the town. Although children were warned not to speak to these women or visit their houses, the madams and their employees were not prohibited from shopping in the general stores or eating in the restaurants. Edna Wilkerson Timmons remembered that the women would patronize her mother's restaurant annually for Christmas Eve dinner and that Silvertip and Lois Landis would leave silk stockings as a gift for Edna. No one could recall any of the women being arrested for prostitution. However, Nona Trembath noted that some of them were arrested on one occasion during the first boom period. Their crime: wearing men's clothing!

These photos were taken in the red-light district, which was located where the Unger house now stands. The woman in the black hat is Lois Landis, the man is probably Jack Landis, and the other women are unidentified. (Photos courtesy of Friends of Midas)

Newspapers

Throughout its history, the town has been home to three local newspapers—two in 1908 and one in 1914. *The Gold Circle Miner* was begun on March 14, 1908, by Mark W. Musgrove, former mining editor for the *Nevada State Journal* in Reno. On that date, Bell Macy wrote in her Day Book: "We got the Gold Circle paper today, the first one that was ever printed. It's a peach." The Gold Circle Publishing Company, incorporated on April 9, 1908, issued the weekly paper for 10 cents a copy. It lasted only a short time and was apparently discontinued that same spring.

Frank L. Reber issued the second paper, *The Gold Circle News*, under the auspices of the Gold Circle News Publishing Company. Beginning June 13, 1908, the paper was an eight-page weekly, distributed every Saturday. Optimistically, subscriptions for the paper were offered for $4 annually. This paper, too, did not last a year and was discontinued sometime during the fall of 1908.

In 1914, Midas gained a registered pharmacist, F.J. Benneson, who not only operated a drug store, but also provided Midas with its third and final paper. Editor, publisher, and "Assisted by an Able Corps of Mining and Society Experts," Benneson produced a handwritten, two-page report of local news and social activities that was mimeographed on old ledger paper. Only a few issues of the weekly *Gold Circle Porcupine* were printed in 1914, but the paper is immortalized as the last known handwritten paper in Nevada.

The *Gold Circle Porcupine* certainly fascinated Elko. In May 1914, the *Daily Free Press* ran a front page story with the headline: "Gold Circle Has A Freak Newspaper." Remarking that the *Porcupine* was "a newspaper that is certainly out of the ordinary and

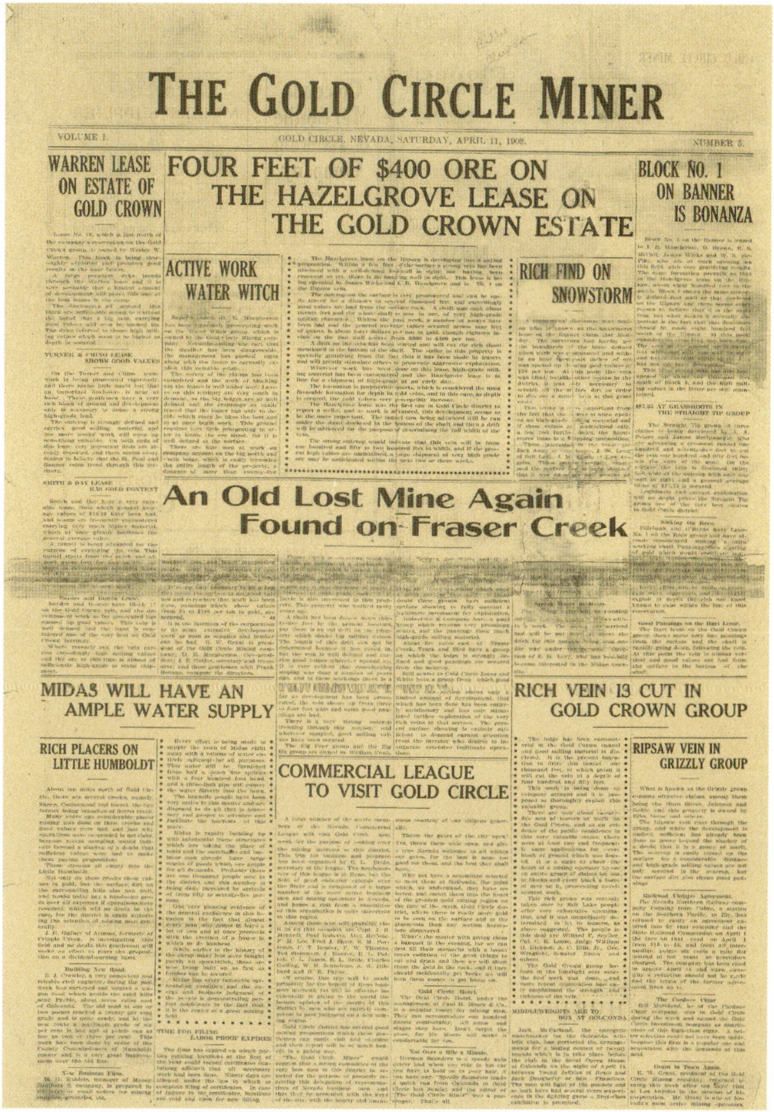
THE GOLD CIRCLE MINER

VOLUME I. GOLD CIRCLE, NEVADA, SATURDAY, APRIL 11, 1908. NUMBER 5.

WARREN LEASE ON ESTATE OF GOLD CROWN

FOUR FEET OF $400 ORE ON THE HAZELGROVE LEASE ON THE GOLD CROWN ESTATE

BLOCK NO. 1 ON BANNER IS BONANZA

ACTIVE WORK WATER WITCH

RICH FIND ON SNOWSTORM

An Old Lost Mine Again Found on Fraser Creek

MIDAS WILL HAVE AN AMPLE WATER SUPPLY

RICH VEIN IS CUT IN GOLD CROWN GROUP

RICH PLACERS ON LITTLE HUMBOLDT

COMMERCIAL LEAGUE TO VISIT GOLD CIRCLE

RIPSAW VEIN IN GRIZZLY GROUP

(Courtesy of the Nevada Historical Society)

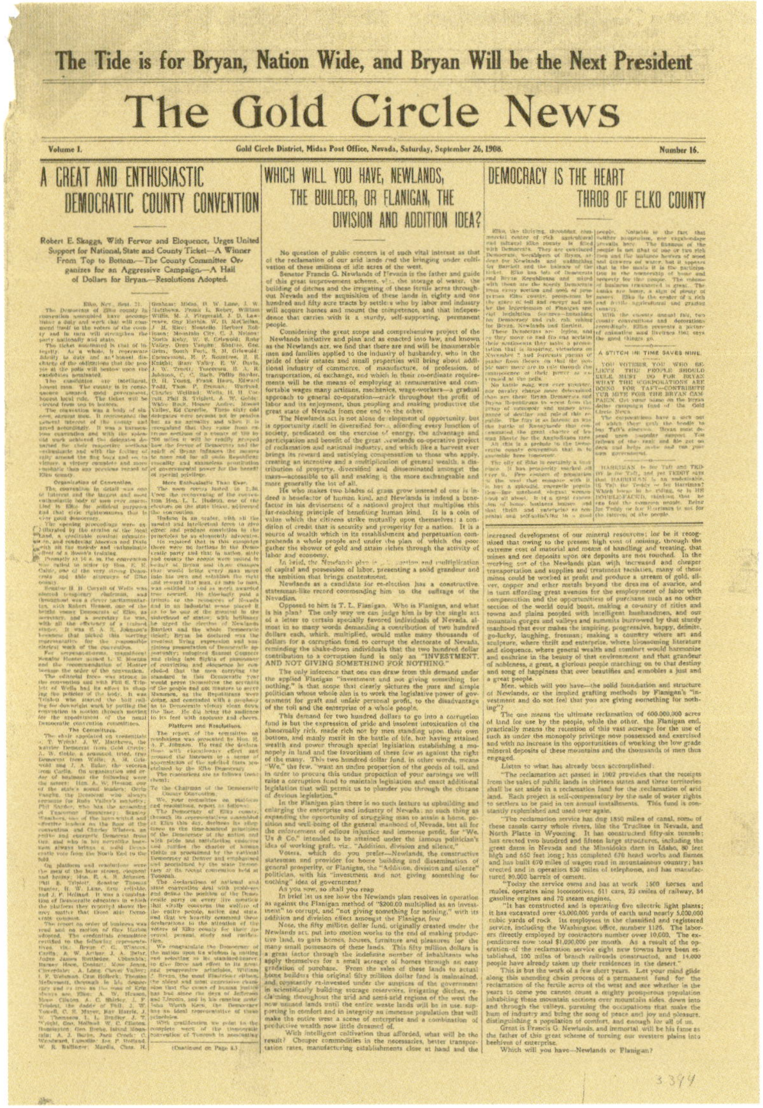
The Tide is for Bryan, Nation Wide, and Bryan Will be the Next President

The Gold Circle News

Volume I. Gold Circle District, Midas Post Office, Nevada, Saturday, September 26, 1908. Number 16.

A GREAT AND ENTHUSIASTIC DEMOCRATIC COUNTY CONVENTION

Robert E. Skaggs, With Fervor and Eloquence, Urges United Support for National, State and County Ticket—A Winner From Top to Bottom.—The County Committee Organizes for an Aggressive Campaign.—A Hail of Dollars for Bryan.—Resolutions Adopted.

WHICH WILL YOU HAVE, NEWLANDS, THE BUILDER, OR FLANIGAN, THE DIVISION AND ADDITION IDEA?

No question of public concern is of such vital interest as that of the reclamation of our arid lands and the bringing under cultivation of these millions of idle acres of the west.

Senator Francis G. Newlands of Nevada is the father and guide of this great improvement scheme, viz., the storage of water, the building of ditches and the irrigating of these fertile acres throughout Nevada and the acquisition of these lands in eighty and one hundred and fifty acre tracts by settlers who by labor and industry will acquire homes and mount the competence, and that independence that carries with it a sturdy, self-supporting, permanent people.

DEMOCRACY IS THE HEART THROB OF ELKO COUNTY

A STITCH IN TIME SAVES NINE.

(Courtesy of the Nevada Historical Society)

would attract attention anywhere," the article reprinted many of the news items from the first issue of this paper, which claimed to transmit "telegraphic news received daily over the longest leased barb wire fence in Elko county":

> A moving picture theatre is one of the near possibilities and no doubt would be an assured fact if our camp pay roll justifies.
>
> The Gold Circle brass band, altho of number few, are still able to render their sweet strains. They do not claim to be professionals, if they were they might at least have the honor of presenting their talents in a 5-cent movie on Market Street, San Francisco. So they have hope.
>
> One of our citizens recently remarked that a 5-cent movie would look like grand opera just now.

The only issue of the *Porcupine* known to have survived the years is dated May 20, 1914. This paper consists of short news notes that mention the work being done at the Sleeping Beauty, the Elko Prince, the Rex, and other mines; the travel plans of many of the town's residents, including the expected arrival of Dr. Mabel Young with her "new auto"; the admirable ability of the "popular school teacher," Miss Violet Shephard, to handle Carl Hammond's "very spirited horse"; and advertisements for only one business: Benneson's Drug Store. According to the notices sprinkled throughout the paper, Benneson's could cure a cold in 24 hours; sold ice cream twice a week in addition to salted peanuts and sweet chocolates; and carried "a full line of the latest remedies" as well as "a full line of up to date popular priced stationery."

Other short articles pointed out "more gardens in Midas this year than ever before," including Al Geabbeart's acre of spuds and John Boyle's "extensive farming at the foot of the canyon." With tongue firmly

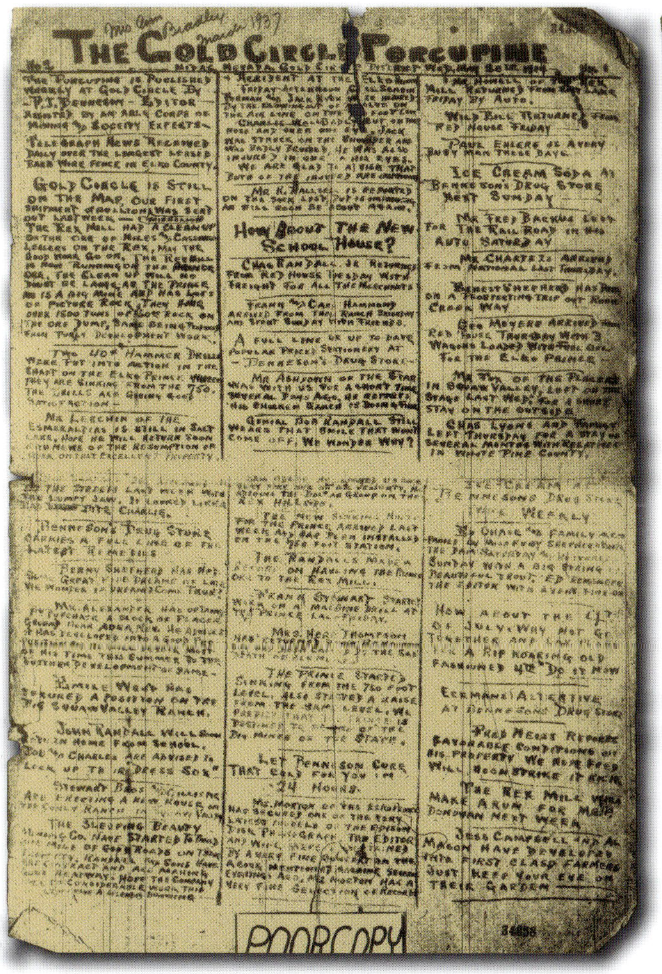

THE GOLD CIRCLE PORCUPINE

MIDAS, NEVADA

GOLD CIRCLE IS STILL ON THE MAP

HOW ABOUT THE NEW SCHOOL HOUSE?

ICE CREAM SODA AT BENNESON'S DRUG STORE NEXT SUNDAY

POORCOPY

"23"

POORCOPY

The Gold Circle Porcupine was the last handwritten community newspaper in Nevada. (Courtesy of Nevada Newspapers Collection, University of Nevada Las Vegas Libraries)

The Midas Band in 1912. From left to right are Charles Randall, Frank Macy, John Randall, Leslie Noble, three unidentified musicians, George Brady, and Bob Randall. (Photo courtesy of Lela Halsell Hempton, Friends of Midas Collection)

in cheek, the *Porcupine* reported that "our town council we understand have contracted for a hitching post in Central Park. Above mentioned post will be provided with a number of big rings to which you can hitch your favorite BULL." Finally, obviously referring to the Elko paper's front page story, the *Porcupine* noted that "several of the state's newspapers have seen fit to notice our humble efforts to boost our camp and have published comments for which we are thankful."

Health Care Services

During its first 20 years, Midas rarely lacked health care providers. Dr. Edward Dunscomb, one of the town's first physicians, moved from Tuscarora to Midas in 1907 to practice medicine and seek his fortune. After his plans for the Dunscomb Townsite faded, Dr. Dunscomb apparently contented himself with being a physician. He delivered several new Midas residents, attended to the town's deceased, and cared for people at the various stages between life's major markers. Two of the babies he delivered were his great-niece and great-nephew, Desda and Gordon Warren born in Midas in 1912 and 1913, respectively.

However, he did not deliver Minerva Macy, according to her mother's day book. Bell Macy's journal did not reveal that she was pregnant until the baby was born at three o'clock in the morning on July 18, 1908. The baby's father, Frank Macy, wrote that day's entry in which he did not mention Dr. Dunscomb. Minerva had the distinction of being the first baby born in Midas. She was another first in 1927 when the first Elko County marriage license of the year was issued to her and Charles G. Lyon.

Dr. Edward Dunscomb (Photo courtesy of Desde Warren Wood, Friends of Midas Collection)

Dr. Dunscomb practiced medicine, located mining claims, platted the Town of Dunscomb, and advocated the creation of a new Nevada county, with Midas as the county seat, until he died in 1915 at the age of 77. He was buried in the Midas Cemetery.

In December 1926, the *Battle Mountain Scout* reported that Dr. Daniel Hiram Pettingill returned to Midas under contract to the Gold Circle Consolidated to staff the company hospital. Apparently, Dr. Pettingill had begun practicing medicine when he opened for business in Midas over two decades earlier, but had been out of the area for an unspecified length of time. His practice ended in the same place it had begun over 20 years earlier when he succumbed to pneumonia on December 8, 1928. That winter was difficult for Midas residents: 25 cases of

Dr. Mabel Young presented this graduation picture of herself to her sister Genevieve Young. (Photo courtesy of BJ Matthews, Friends of Midas Collection)

Hamming it up for the camera in Midas in 1911 were (left to right) Dr. Mabel Young, "Jennie's sister," an unidentified woman, and Mabel's niece Genevieve Young (known as Jennie or Gen). (Photo courtesy of BJ Matthews, Friends of Midas Collection)

influenza were reported, including his own. It was a bad time to be without a doctor. He was the last doctor to maintain a practice in Midas. His widow, Ethel, was still living in Midas two years later, according to the 1930 census.

The town also briefly boasted a small hospital. *The Battle Mountain Herald and Central Nevadan* announced on April 23, 1908, that William Menken, Carl Schein, and Charles Dorsey had just completed building a hospital in Midas that could care for up to eight patients at one time.

Another professional woman conducting business in the new district was Dr. Mabel Katherine Young, one of Nevada's first female dentists. Dr. Young was born in Wisconsin in 1882 and first visited Elko County when she was seven years old. In September 1889, the Elko *Weekly Independent* lamented her return to Milwaukee after having lived with her brothers William, Richard, and Dean on Van Duzer Creek, noting that "Miss Mabel was exceedingly popular among a wide circle of friends in Mountain City and Tuscarora."

She graduated from the Dental Department of the Milwaukee Medical College (one of two women in a class of 50) in 1904 and moved to Colorado. It is possible that she may have met Dr. Dunscomb in Cripple Creek as they were both there in 1904.

By 1906, Dr. Young had arrived in Elko to open her first dental office. The Elko *Free Press* explained that she was also a post graduate of the Chicago Northwestern Dental Surgery College and noted that Dr. Young "passed such a splendid examination before the [Colorado] State [Dental] Board that Governor McDonald of Colorado wrote Governor Sparks [of Nevada] requesting him to issue a permit for her to practice in Nevada until the meeting of the State Dental Board, when she will take the examination."

When the rush to Midas began in earnest in 1908, Dr. Young was among the fortune-seekers. Other family members also made their homes early in Midas, including Luella

Pendergast who had once been married to Dr. Young's brother William; Dr. Young's sister, Genevieve, who was married for some years to long-time Midas resident Herbert A. Thompson; and her niece, Genevieve Young. According to the 1910 census, Dr. Young and her niece were living with her sister and brother-in-law, the Thompsons.

Dr. Young immediately opened a dental office on North Main Street in Midas, but she certainly was not relying on the residents' sore mouths to make her fortune. Although the 1910 census noted that dentistry was her stated occupation, she was also the town's third postmaster, serving from July 16, 1909 until August 13, 1911. In addition, this remarkable and busy woman was the manager of the Dunscomb Townsite Company and the town's correspondent to the Elko *Weekly Independent*, writing "Gold Circle Items" in 1908 and parts of 1909.

As the president and manager of the Dunscomb Townsite Company, which sought to sell lots and establish a town near the Esmeralda Mine, Dr. Young actively advertised the town in 1909. Although the Elko County Commissioners approved the plat map for the new town, it is not known if any lots were ever sold.

In 1912, Dr. Young began traveling by horse and buckboard to nearby mining camps and ranches to provide dental services. A year later, Josie Alma Woods, future Eureka County Assemblywoman, joined Dr. Young as her assistant and companion in her itinerant practice.

Two years later, they had traded in the horse, and the residents of Midas waited with great anticipation for the arrival of Dr. Young in her new car. Many years later, Woods recalled that she paid a mechanic $50 to take that car's engine apart and show her how everything worked. She wanted to be able to fix any problem they might encounter on the rough Nevada roads. Considering that Dr. Young's dental fees ranged from $1 for a cleaning to $10 for a single crown, $50 was a considerable sum.

The only man in this picture is actually dressed as a woman! It is not known why this picture was taken; however, Dr. Young liked to write and produce plays. Perhaps this is the cast for a Midas show. On the left is "Gen's sister" in the light-colored shirt, tie, and pants. To the right, loosely making up the back row, are Mrs. Savage, an unnamed female visitor, and "Piff," a girl visiting from Golconda. In the front row, from left to right, are Dr. Young (in the dark suit with hand on hip), her niece Gen Young, and Mr. Savage in Mrs. Savage's clothes. (Photo courtesy of BJ Matthews, Friends of Midas Collection)

Dr. Mabel Young in Midas in 1911. (Photo courtesy of BJ Matthews, Friends of Midas Collection)

By 1915, the women were traveling an itinerant dental route that included, primarily, Austin, Eureka, Round Mountain, Manhattan, and Tonopah. One of her Eureka patients recalled that Dr. Young, a small woman (about five feet tall), sometimes sat in the laps of her patients in order to extract their teeth. By 1917, it appears that she had established residency in Eureka as she had bought a home and office in that town. In 1922, Dr. Young ran an unsuccessful campaign for Eureka County Clerk and Treasurer. She was the first woman to run for that office.

In 1927, the two women bought a ranch outside of Eureka in Grass Valley, but Dr. Young would not live there long. She became ill, and in 1933, she deeded the property to Woods and moved to Los Angeles, California. She died there in 1939.

Currently, no practicing health care provider has a Midas-based business. However, the nearby presence of Newmont's Midas Mine ensures that emergency medical services are available to town residents should they be needed.

Transportation

Gold-seekers and others looking to make their fortunes in Midas arrived in the new mining district by various means and from different locations. Some people were transported by auto- or horse-drawn stages; a few had personal automobiles. Most traveled by horse or horse-drawn wagons. The only existing roads came from Battle Mountain, Tuscarora, and Golconda. For the most part, these roads were not in good shape.

The district was accessible primarily from Winnemucca and Golconda by the existing Squaw Valley Ranch road. Enthusiasts in Winnemucca talked of improving the road so that the time needed to travel over 60 miles between their town and the new camp would be reduced to three hours.

In October 1907, the newly-organized Battle Mountain Commercial League announced that one of its first projects would be to shorten the road between Battle Mountain and Gold Circle. The league recognized that Battle Mountain could become the major supply center for the new camp because it was the closest railroad connection. Completed in April 1908, the new road shortened the distance between Battle Mountain and Midas by 12 miles.

Traveling the 90 miles between Elko and Gold Circle was not as easy. The Tuscarora paper scolded Elko County in February 1908 for the condition of the road between Midas and Tuscarora. It was so bad that mail had not been moving between the two towns. The paper exhorted Elko to improve the road or risk losing Midas business to Golconda or Battle Mountain. The *Free Press* editor agreed with his Tuscarora colleague and recommended that the county quickly remedy the situation. The county, apparently, did not respond for 15 years.

On August 22, 1923, the *Nevada State Journal* reported that "a much needed road ... is being constructed from Independence valley over an old abandoned route to Midas by the county road crew" and should be completed by autumn. Noting that travelers between Elko and Midas were required "to

George Meyers freighted the 24-ton diesel engine to the Elko Prince Mill in 1911. He believed it to be the heaviest load ever hauled by a horse-drawn wagon in Nevada. (Photo courtesy of Lela Halsell Hempton, Friends of Midas Collection)

This is probably George Meyers freighting in the winter of 1915. (Photo courtesy of Lela Halsell Hempton, Friends of Midas Collection)

go around by Golconda," the article explained that:

> The new road will follow an old wagon road, except for a short distance, where it will be routed through the Willow Creek canyon to eliminate the heavy grade which has made travel almost impossible.

Despite the new road, traveling between Midas and Elko continued to be as difficult as it had always been, much to the annoyance of Midas residents. This inability to get to their county seat led several to advocate removing Midas from Elko County. In 1921, Midas residents asked the Nevada Legislature to redraw the county lines in order to place the town within the borders of Humboldt County. Undaunted by the lack of legislative interest, a similar petition was circulated two years later by Jerry Coughlan, Superintendent of Big Chief Consolidated Mining. He explained that the road to Elko was worthless seven months out of the year "due to lack of bridges and impassible mountains." In 1927, an allegation was printed in the Elko newspaper that Humboldt County legislators would introduce such a bill; the paper later noted that the story was not based on any fact. Although complaining about the county government appears to be a long-standing tradition in Midas, the town has remained in Elko County.

However, the primary route between Midas and the rest of the world begins in Humboldt County. For most of its existence, the condition of that gravel road, commonly known as the County Road, was dependent on sporadic government maintenance, but it was always in better shape than the dirt road from Midas to Tuscarora. The road from Battle Mountain had been virtually abandoned. There is no paved road within 20 miles of Midas.

In 1997, Midas Joint Venture entered into an agreement with Elko County to maintain the part of SR 18 that runs from the Humboldt County line to the mine road, which is about one mile past the turn-off to Midas. Subsequent Midas Mine owners, Normandy and Newmont, have continued to grade and treat the road. Although the mining companies sought to improve the road primarily for the buses that transport employees and the trucks that transport equipment, they have also provided a tremendous benefit for the residents and visitors of Midas. For 90 years, the condition of the county road was a low priority for local governments; now, travel into and out of Midas has improved exponentially. Not only do mine employees maintain the county road, they also maintain the two-mile stretch into town, plowing away the snow in the winter and treating it with magnesium chloride in the summer. Mine workers also repair the road more quickly than public workers ever did, as was evidenced when parts of the route into Midas washed out in the spring of 2005 and in January of 2006.

Currently, there is no public transportation into Midas, but in the days before everyone had his or her own pickup truck, public transportation was essential. The first was the Elko-Tuscarora stage, which added a run to Gold Circle in October 1907. In February 1908, Charles Dorsey began a stage line between Battle Mountain and Gold Circle. The article announcing the new business claimed that "passengers can now be taken through in one day." C.H. Snow started his auto line between the two towns in the spring of 1908. He advertised that his cars could travel the 40 miles to Gold Circle on the new road in three and one-half hours for $15 one way or $25 round trip. By April, business was sufficient for Snow to utilize three automobiles in his fleet and to order a fourth.

This load of freight arrived in Midas by steam engine in 1919. The false-front building without a sign (next to the Primeaux store) was used for, among other things, a land office and a bar. (Photo courtesy of the Nevada Historical Society)

Traveling between Midas and Golconda around 1918. (Photo courtesy of Len Berg, Friends of Midas Collection)

Louis Reymond waiting for his next passenger. It is not known if this picture was taken near Midas. (Photo courtesy of the Lyon Family, Friends of Midas Collection)

In March 1908, an auto run was begun between Winnemucca and Gold Circle. The service consisted of a nine-passenger car that came from Fallon. The car was originally delivered to Gold Circle after a "quick" nine-hour trip from Fallon. Throughout much of 1914, the *Reno Evening Gazette* carried a classified advertisement under the category "stage lines" for a daily "auto line" between Midas and Golconda, operated by W.L. Brackett. The ad promised "new, large, roomy cars always in service."

Because of its isolation, the town relied heavily upon a strong transportation system. In the early years, auto and stage lines ran daily between Midas and the neighboring towns of Golconda (a Thomas Flyer carried passengers) and Battle Mountain, transporting fortune seekers to the district. Freight wagons, usually pulled by horses or mules, later by gasoline engines, also traveled frequently to Midas, bringing needed supplies to the growing town.

Despite the numerous opportunities for surface transportation to Midas, these were not fast enough for some people. On November 11, 1909, Dick Keyworth and Charley Burns of Battle Mountain announced their plans to provide a "flying machine" service to three of the mining districts closest to Battle Mountain, including Gold Circle. They explained to their local newspaper that a machine had been ordered from the Orville Wright Company with the backing of Nevada mining millionaire John Harmon. The service was to begin on February 2, 1910, according to the enthusiastic account. Travel time between Battle Mountain and Midas was expected to be 22 minutes. The paper noted that:

> A speedier schedule could be arranged but the whirlwind driver, Dick Keyworth, says, "That will be fast enough for

the swiftest of them."

Later, the planners announced that the order had been changed to a "Darland Heliceptor" [sic] from San Francisco, but the start date remained the same. The people scheduled to fly the first circuit in the four-passenger vehicle were pilot Keyworth, George W. Abel, Frank L. Reber (by then a former Midas resident), Tom C. Parker of the Battle Mountain paper, and backer Harmon. But the date for the beginning of the service came and went without any comment from the paper or the promoters. Obviously, the grandiose plan had been abandoned.

The failure of the air service plans did not deter other dreamers. Experienced mining people knew that the proximity of a railroad could make or break a district. Although Midas was closer to the railroad, via Battle Mountain or Golconda, than were many other Nevada mining towns, the cost of transporting ore to those connecting points was prohibitive. Consequently, plans to build a spur into Midas materialized early.

The 1909 Nevada Legislature and Acting Governor Denver S. Dickerson approved two bills concerning the building of a railroad to Midas. Senate Bill 35, introduced by Senator William Easton of Lander County, was approved on March 11, 1909, and granted the right of way to Jacob Eggers, C.B. Welshons, E.L. Dutertre, and John I. Peterson for the construction of a railroad from Golconda to the northern boundary of Elko County, passing through Midas, Burner Mining District, Lime Mountain, Edgemont, and White Rock. Assembly Bill 162, introduced by Assemblyman F.H. Winter from Elko County, was approved two days later and granted a right of way to C.G. Rothchild, F.L. (Frank) Reber, E.E. Oldham, E.A.P. (Edgar) Johnson, and A.B. Gray to build a railroad from Midas through Tuscarora to a point in Elko County on the Western Pacific or Southern Pacific line.

The first right-of-way was granted for a term of 50 years and the second for 25 years, provided that construction begin within 18 months and be completed within 10 years from the passage of the legislation. Neither measure contained an appropriation from the state. Needless to say, the railroad, like the flying machine service, did not come to Midas, despite the most enthusiastic of plans.

Fortune Beckons to the Wealth-Laden Hills

of

Gold Circle Nevada

Guide and Manual

The World's Best Crop

Its price never falls
It never deteriorates
Its market is world-wide
It cannot be over-produced
It pays every debt
It buys every commodity
It banishes poverty
It employs labor
It stimulates industry
It rewards courage

It Is

GOLD

Issued by

CHAMBER OF COMMERCE
Midas, Elko County, Nevada
Copies on Request

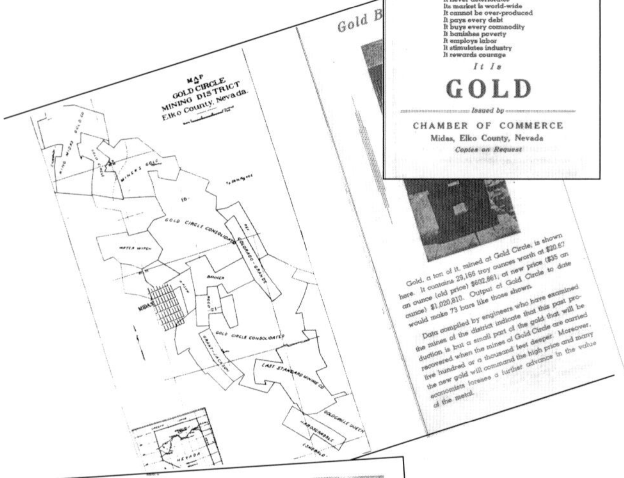

Gold B

Gold, a ton of it, mined at Gold Circle, is shown here. It contains 29,166 troy ounces worth at $20.67 an ounce (old price) $602,861; at new price ($35 an ounce) $1,020,810. Output of Gold Circle to date would make 73 bars like those shown.

Data compiled by engineers who have examined the mines of the district indicate that this past production is but a small part of the gold that will be recovered when the mines of Gold Circle are carried five hundred or a thousand feet deeper. Moreover, the new gold will command the high price and many economists foresee a further advance in the value of the metal.

Hardscrabble—Mine: Hardscrabble. Mrs. Rice, New York City, owner. Claims, 3. Deepest development, 200 feet. Production, $50,000.

Champion Gold Mining Co.—Mine: Champion. Claims, 4. Development, 600 foot tunnel. Production, bins full of ore. Operated by Noble H. Getchel, Jack Dempsey and associates.

Mills of Gold Circle District

Cyanide Mill of Gold Circle Consolidated. Capacity 75 tons a day.

Amalgamation mill of East Standard Mining Company. Capacity 60 tons.

OTHER MILLS

Amalgamation mill of Esmeralda mine. Capacity 15 tons.

A 25-ton flotation plant is projected for the Miners Gold and Midas mines.

Young Custom Mill.

So Investors May Know

Total production of gold and silver to the value of nearly $5,000,000 by eleven mines in the Gold Circle district is shown by the data available. Gold was paid for at $20 per ounce and silver at the average of 40 cents per ounce. The same production now, with gold at $35 and silver at 64½ cents would approximate $8,000,000, a gain of $3,000,000 for the producers.

The deepest mine in Gold Circle district is down 900 feet, the average depth of the others is around 200 feet. In the deep mine some of the best ore bodies occur below the 200 and continue to the 900-foot level where values are higher than above. Experts find no geological obstacles to the continuance of ore down to and below the 900-foot level in all mines.

While the values of gold and silver have risen 69 and 61 per cent, respectively, the production cost of the metals in Gold Circle district is practically the same as in 1916, leaving a broad margin for profit.

Gold Circle has attracted many prominent figures in the mining world. Among those who have invested money in the district are Charles A. Stoneham, millionaire mining broker, now owner of the New York Giants ball club; Noble H. Getchel, Nevada state senator and veteran mine operator; Harry S. Joseph, Utah mine developer; W. F. Snyder & Sons Company, associated with the National Lead Co.; George Wingfield, Nevada mine operator and banker, and others. Jack Dempsey, late heavyweight champion, who was reared in the mining country, has invested some of his fortune in Gold Circle.

Chambers of Commerce

Midas businesses and mines have, over the years, enjoyed two organizations whose primary purpose was to attract investors into the district.

The first, the Gold Circle Commercial Club, was formed on March 15, 1908. Similar to a chamber of commerce, the Club held its first meeting in "Dorsey's hall" that evening. From the time the membership list was opened at 10:00 that morning until the meeting was convened, 66 people paid their initiation dues of $50. The Club, gathered under the direction of Clarence Rothchild and S.L. Cory, decided to rent, for $75 a month, Ulrich hall on Main Street for its headquarters and that monthly dues would be $5. *The Gold Circle News* printed the Club's motto on the front page of the June 20 issue:

> A commercial and social club, so as to help us in helping each other, to make this a well advertised district and letting the outside world know that WE HAVE THE GOLD HERE AND PLENTY OF IT.

The paper went on to identify the elected officers: President, H.H. Riddels, "the progressive business man"; Vice-President, A.A. Primeaux, "a princely merchant"; Secretary, J.W. Saulter; and Treasurer, Frank M. Shirer. The Board of Directors consisted of J.W. Matthews, J.H. Moffatt, H.W. Lane, William Menken, S.M. Levy, R.H. Grieve, P.A. Leamy, and Clarence Rothchild, "all men of force, character and standing in the commercial and mining avenues of the west."

The Gold Circle Commercial Club organized some of the earliest town celebrations, such as Independence Day and Labor Day, but the group does not seem to have lasted long.

In 1932, a public meeting was held at the Town Hall to organize a new Chamber of Commerce for the town. The stated goal for the group was to "advance . . . the interests of the Gold Circle Mining District." Elected officers were Noble H. Getchell, President; Charles S. Warren, Vice-President; J. A. Barclay, Secretary; and Charles B. Lyon, Assistant Secretary. Additional board members were Mike J. Paisley and Victor Jacobson.

Chamber activities varied. In July 1933, the organization sent a telegram to U.S. President Franklin D. Roosevelt requesting the lifting of the embargo that prohibited the shipping of gold to other countries. The Chamber also produced a promotional brochure titled *Gold Circle Nevada: Guide and Manual.* Claiming that "Fortune Beckons to the Wealth-Laden Hills," the brochure included a map of the district's mines and pictures of Midas, gold ore mined in the district, and three mills. It provided directions for traveling to the district by automobile, railroad, and airplane, noting that accommodations were available at a hotel and boarding houses. It also named all of the active mining companies and summarized their past production and future plans.

To entice new capitalists to the area, the brochure listed some of the current investors in the district:

> Charles A. Stoneham, millionaire mining broker, now owner of the New York Giants ball club; Noble H. Getchell, Nevada state senator and veteran mine operator; Harry S. Joseph, Utah mine developer; W. F. Snyder & Sons Company, associated with the National Lead Co.; George Wingfield, Nevada mine operator and banker; and . . . Jack Dempsey, late heavyweight champion.

Potential financiers were assured that "while the values of gold and silver have risen 69 and 61 per cent, respectively, the production cost of the metals in Gold Circle district is practically the same as in 1916, leaving a broad margin for profit." Finally, the brochure revealed that "data compiled by engineers who have examined the mines of the district indicate that this past production is but a small part of the gold that will be recovered when the mines of Gold Circle are carried five hundred or a thousand feet deeper." At that time, the deepest mine, the Elko Prince, was down 900 feet. Other mines averaged a depth of 200 feet.

Later in 1933, the Chamber sponsored a Labor Day celebration. The party was organized by Harry S. Joseph, John J. Hanley, and Charles S. Warren of the Miners Gold Company. The article in the *Battle Mountain* Scout contained the following account:

> Local residents and numerous visitors enjoyed one of the biggest celebrations in the history of this mining camp . . . Streets were decorated in a gala array of bright colored flags and balloons and the use of confetti during the day caused considerable hilarity. The celebration was officially opened by a salute by bombs at 7:00 o'clock in the morning and the entire day was a continuous round of sport and pleasure, for the children as well as the grown-ups.

The schedule for the day included games and contests in the morning; a "public lunch" at noon; and horse races and numerous bronco busting, roping, and bull-dogging contests in the afternoon. The Midas baseball team defeated the visiting team from the Elko Industrial School, and "the celebration closed with a grand ball in the evening."

The Midas baseball team played on its own diamond at the Squaw Valley Ranch outside of town. (Photo courtesy of Northeastern Nevada Museum)

This baseball logo was found on a 1909 letter written by Frank Reber. (Courtesy of Friends of Midas)

MIDAS SOCIAL LIFE

Stories have survived that indicate the residents of Midas were not always a serious lot. One example of frivolity in the town occurred in 1910 when a Midas resident played a huge practical joke on his fellow citizens. The Elko *Free Press* reported that Dick Avery woke everyone in town at 3:30 a.m. on April 27, 1910, to see Halley's Comet, which was expected that year, appear over the hill to the east of Midas. Apparently, he duped many of the residents into believing that they had seen the famous comet.

Two days later, the paper noted, Avery "let the cat out of the bag" and produced a 30-foot pole that had had a fiery, oil-soaked bundle at one end with which he had dashed headlong down a hill overlooking the town. The paper explained that not everyone found the prank amusing:

> Now that it is known that it was all a josh, some of the women that walked a half a mile or more (to get a better view of it) are very sore about it (the josh).

Apparently, however, such elaborate events rarely occurred during this period in Midas. In 1912, the Midas reporter to the Elko paper commented that two occurrences had caused a "ripple" in the "usual quiet calm of our social life": the departure of one of the town's boosters, James Boyle, and the divorce of an unidentified "popular young married couple."

On June 14, 1916, the Midas columnist for the Elko newspaper provided one of the few accounts of a wedding celebration in the town. The reporter recounted the surprise of the residents when Delia Hart and Kurt Cornwell returned home, newly married, on June 6, and detailed that evening's festivities:

> An impromptu party was held at the home of the bride's parents. Everybody had a hilarious time and well wishes were showered upon the happy couple. The next evening they were serenaded, as the evidence of battered coal oil and gasoline cans which littered the street next morning evidenced. We are not so busy chasing the almighty dollar here that a social event like this does not liven us up some.

However, the town "livened up" often that summer. The correspondent reported "a moving picture show and a dance to follow every Saturday night now."

According to regional newspapers, the women of Midas were active in charitable activities relating to the involvement of the United States in World War I. On April 17, 1918, the Elko *Independent* observed that the Midas district of the Woman's National Liberty Loan Committee was one of the most successful in Elko County. Chaired by Mrs. R.T.(Mary) Noble, the group had raised $6,000, exceeding its goal by $1,000. Of the 28 districts in the county, only Elko and Wells surpassed Midas in fundraising. Six months later, the Elko *Daily Free Press* noted that every resident of Midas "has taken a liberty bond." The paper asked its readers: "Can Elko equal this record?"

The newspaper also reported on the Midas chapter of the Red Cross. Organized on January 1, 1918, the chapter had grown from 15 to 25 active members. The chapter's officers were Mrs. R.T. (Mary) Noble, Chairman; Mrs. A. (Ellen) Primeaux, Secretary; Mrs. J. (Mamie) Cody, Treasurer; and Mrs. W. (Agnes) Clawson, Chairman of the Workroom. The members of the chapter met two times each week to work and two times each month to conduct business.

The paper praised the Midas chapter, boasting that "the noble and self-sacrificing spirit of the ladies can best be illustrated by the fact that in six months the following was shipped to the headquarters: 290 Hospital Garments, [unreadable], 23 sweaters, 23 pairs of socks, 3 helmets, 500 pillows and many wash rags from school children." The chapter also organized the Fourth of July celebration for 1918, with the proceeds going to the Red Cross Fund. The activities during that holiday included the dedication and raising of the honor flag awarded to the town for twice exceeding its goal in the Liberty Bond drive.

On November 11, 1918, Nevada's Acting Governor Maurice J. Sullivan declared a state holiday to celebrate the end of the "War to End All Wars." In Midas, the armistice was observed with a parade down Main Street and rowdy festivities.

Two years later, the residents were planning thespian activities. According to the *Nevada State Journal*:

> Under the supervision of Mrs. John Boyle and Mrs. W.T. Hargrove a play will be staged at the Midas auditorium some evening next month [June 1920], with local talent in the cast. The following are those who will take part: Mesdames Ben Meyers, John Cody, W.T. Hargrove, Leslie Noble, Fred Holainger; Misses Minerva Macy and Lucille Parkinson, Judge Frank Macy, Jerry Coughlan, Jess Campbell and "Doc" Brinker.

Former residents and past issues of the Battle Mountain *Scout* provided information about the social activities of the town's residents in the 1930s. The Battle Mountain and Winnemucca newspapers, particularly after their merger in 1935, often printed a column that detailed the activities of Midas citizens. The column debuted in July 1926 and continued sporadically through 1942. Usually titled "Midas News Notes," the column's author appears to have been Mrs. Bernard (Agnes) Gilmour from 1929 to the early 1930s and Mrs. Isabel Lyon from 1934 to 1942; however, a byline was not always printed. The column chronicled the arrivals, departures, illnesses, weddings, parties, and other activities of Midas residents and visitors. It reported on entertainment conducted in town; the plans for church services; and the participation of Midas residents in Masonic lodges and Eastern Star chapters in Tuscarora, Battle Mountain, and Winnemucca. "Midas News Notes" also followed the accomplishments of the town's residents. In April 1933, the column reported this success of a 24-year-old who had grown up with the town:

> The song, "She's All For Me," composed by Don Hargrove of Midas, was sung over station KTFD, Twin Falls, Idaho, by H. Bates, April 4, at 8:30 p.m. Many Midas people heard it.

"Stair Steps" Don Hargrove, George Myers, Arvid Jossis, Sig Jossis, 1930 (Photo courtesy of Anna Jossis Hinrichs, Friends of Midas Collection)

Desda Warren and her brother Gordon pose next to the family's home on Main Street in Midas in 1919. (Photo courtesy of Desda Warren Wood, Friends of Midas Collection)

The Midas Town Hall as it appeared in the 1960s. (Photo courtesy of Edna Wilkerson Timmons, Friends of Midas Collection)

The Town Hall can be seen behind Don Richardson and his unidentified friend. (Photo courtesy of Don Richardson, Friends of Midas Collection)

Mary Wilkerson explores the interior of the Town Hall after it was no longer actively used. The planks on sawhorses along the wall provided a stage for performances. (Photo courtesy of Edna Wilkerson Timmons, Friends of Midas Collection)

The Midas Town Hall in the late 1970s not long before it was torn down. (Photo courtesy of Friends of Midas)

Town Hall

To accommodate the many social and community activities, a town hall was built sometime in the early 1920s. Ray Clawson, who was superintendent of the Elko Prince Mill in 1922, recalled that he and his father, William H. Clawson, installed the hardwood floor in the building that was actively used for many years. Former residents fondly recalled numerous dances, chivarees, and card parties at the hall. Midas women were particularly fond of cards. Throughout this period, the column mentioned several regular bridge clubs and reported on various card parties at residents' homes. For example, in 1938 the formation of another bridge club was announced. It consisted of 12 women who planned to play twice a month. The Midas column also referred frequently to dances at the Town Hall for any occasion — holidays, benefits, to honor newlyweds, waiting for election returns — or for no specific occasion at all. Music was usually live and provided by the "Midas Orchestra" or the "Idaho Night Hawks." In 1928, one of the dances featured Midas residents Laura and Walt Landon on the piano and drums and W.E. Cody on the saxophone. At least once, just a radio provided the music for the dancing enthusiasts of Midas.

The Midas Town Hall was also the site of school and holiday programs, including a ceremony for the graduates of the eighth grade, the highest level usually taught at the Midas School. Additionally, the Town Hall housed the community Christmas tree, a pine imported to Midas each year. The "Midas News Notes" for January 5, 1934, provided the following account of a typical Midas Christmas:

> The Christmas tree and program held at Midas hall December 22 was a grand success. Miss Adele Myers deserves considerable credit as all the children were well-trained. A large crowd attended and all spoke with praise about it. The Sunday school choir also rendered lovely songs.

Two years later, the column reported that "the stage is set and

Astrid Berg and her doll enjoy a stroll down Main Street around 1918. (Photo courtesy of Len Berg, Friends of Midas Collection)

the actors are about ready for their part in the Community Christmas program. The donations toward the children's tree and program have been so generous that Santa Claus will be able to remember every child."

Many of these activities and dances were instigated by the Mothers' Club. Suggested by the town's teachers, the club was formed on October 23, 1926. The first officers were Mrs. R.T. (Mary) Noble, President; Mrs. C.S. (Ruth) Warren, Vice-President; Mrs. A.A. (Ellen) Primeaux, Secretary; and Mrs. S. Tremble, Treasurer. In May 1929, the Mothers' Club held a dance and raffle to pay off the mortgage on the hall. The raffle was for a cedar chest, which was won by Desda Warren. The club raised $526.50, exceeding the $375 debt remaining on the mortgage.

Traveling shows also appeared at the Midas Town Hall. Interestingly, the best source for the names and dates of these troupes was the graffiti on the interior walls of the hall. Before the building was demolished, the numerous carvings were transcribed. The Standard Show Company visited Midas twice in 1928, noting the "Good house nice people." Other companies included the Morries Laffun Fools Road Show, which cited "poor biz" in August 1931, and the Plunketts Minstrels, who played to 15 people in October 1937 and enjoyed "good biz."

In the spring of 1936, the "Midas News Notes" reported that Friday evenings in the town were devoted to "practice dancing" for children, led by Mrs. Ida Fox and Grace Robear. The town's saxophone player, Charles Lyon, assisted at these Friday dances, and adult classes were planned. The town was snowbound, but the plow had arrived, and travel was expected to improve. The columnist assured the outside world that "we of Midas make our own amusement, mostly with dancing and cards. The roads have been too bad to get out elsewhere."

When not dancing or playing cards, the residents formed other social organizations. In September 1936, the columnist indicated that "our young men of Midas are planning an athletic club for the winter." A new Red Cross chapter was formed with Mrs. Ray (Grace) Clawson as the Midas representative. The 4-H club was first mentioned in the "Midas News Notes" in June 1938 when a dance was held at the Town Hall to benefit the club. It was deemed a big success with visiting dancers from Tuscarora, Golconda, and nearby ranches. The next month, the 4-H club presented a play in Midas. Following a popular local run, the play was performed in Wells, with Mrs. Al (Ida) Fox and Mrs. H. F. (Lounella) Wilcox chaperoning the young thespians.

In 1965, *Desert Magazine* reported:

> Travel to Midas today and you find the same buildings, but with changes. The town hall stands empty, but ready for a meeting. The yellowed keys of a rinky-tink piano inside eagerly await the caress of lively fingers. A well battered ballot box awaits duty should someone come to call for an election.

Both items were moved to the Midas Schoolhouse Museum in 1998, and both were lost in the 2005 fire.

Despite many efforts to save the old building, time and neglect took its toll on the Midas Town Hall. It was torn down in 1986 for safety reasons. The boards from the hardwood floor were carried down the street and installed in The Kountry Store, owned at the time by Sam and Edythe Seals.

Desda Warren, Minerva Macy, Jenny Lewis, in 1925 by the side of the Lewis Restaurant (Photo courtesy of Anna Jossis Hinrichs, Friends of Midas Collection)

Titled "Hen Party," this photo shows the denizens of Midas society in the 1930s. From left to right are Mrs. Cynthia Thompson, Mrs. Marion Noble, Mrs. Thelma Clawson, Mrs. Ellen Primeaux, Mrs. Hilda Jacobsen, Mrs. Grace Clawson, and Mrs. Julia Pansar. (Photo courtesy of Anna Jossis Hinrichs, Friends of Midas Collection)

Midas has always loved a Fourth of July parade! The cars decorated at the Elko Prince Mine (where the Berg family and others lived in 1919) make their way down Main Street. (Photo courtesy of Len Berg, Friends of Midas Collection)

Oscar Berg is ready to commemorate the Fourth of July in 1919. (Photo courtesy of Len Berg, Friends of Midas Collection)

The Fourth of July

This holiday has always been a particular favorite in Midas. The first celebration was held in 1908, and each year after that saw parties both elaborate and modest, depending on the town's population at the time.

The first duty undertaken in 1908 by the Gold Circle Commercial Club was planning the coming Fourth of July celebration. A committee was appointed to, as *The Gold Circle News* so eloquently stated:

> arrange for a natal celebration and have the eagle scream with all the fervor and eloquence and patriotism that is ever such a manifestation of the sturdy Americans in these mountain mining districts. Mark the first Fourth of July celebration . . . will not be a tame affair.

The Elko paper printed an article outlining the many activities planned, including team and single hand drilling contests, horse races, mine tours, ladies' races, sack races, fireworks, and a dance. The article was written by "JWM" (probably James W. Matthews), who issued the following invitation to Elko:

> We would be pleased to have you visit our club (there will be no 20-round bouts on that day). You can bring your lemons with you and we will furnish the water from our new water works.

The papers did not report the winners of the various contests, but an interesting story about this particular celebration emerged in the *Nevada State Journal* in 1929:

> Some 20 years ago in Midas two strange double jackers came in and won the Fourth of July drilling contest. Next day they bragged about it and Bill Menken (he is manager of the McCoy property 35 miles south of Battle Mountain now) told them he could beat their record in 10 minutes straight away without changing hammers if he could get Frank Benane to turn for him.
>
> So Frank was hunted up and he said he would do it. He is still in Gold Circle. Part way through the time Bill split a steel and half of it went through Frank's hand, but when Bill asked him if he was hurt he said, "No," and went right on turning until he knew

the hole was deeper than the one the two men had made.

He had measured their hole and marked the steel. Menken had beaten them in eight minutes and he never had drilled in a contest, nor had he ever considered himself much good at drilling. He was a great big fellow then; weighed more than 250 pounds, and he could hit a drill a terrific wallop.

Thus began an important Midas tradition. Despite the fluctuating fortunes of the area, most summers have included festivities to mark the occasion. The contests may have changed, but the fun remained the same.

In 1926, the Battle Mountain *Scout* proclaimed the two-day festival a "jolly and interesting" celebration. The next year, it noted that $1,000 was available to fund the festivities because Gold Circle Consolidated employees had each donated one day's pay. Management, including Noble Getchell, also made donations. These celebrations were busy affairs, with activities such as baseball games, horse races, bucking contests, drilling contests, tug-of-war, kids' games and races, "appropriate exercises in the Town Hall," and, of course, a dance at the end of each day's activities.

Each July during the 1920s and 1930s, the "Midas News Notes" commented about that year's Fourth activities. In 1934, the columnist noted that "Midas celebrated July 4 with a clown parade at 1:30 p.m., after all enjoyed a delicious lunch in the hall, then races of all sorts, a dance at night, fireworks and firecrackers." Often, the Independence Day festivities included games involving the Midas Baseball Club, which had been formed by the residents of the town to compete against teams from towns throughout Elko, Humboldt, and Lander Counties. Gordon Warren, who played on the squad as a young man, recalled that the team had a nice ball field near the entrance gate to the Squaw Valley Ranch.

Not many accounts of the games were printed. Those reports that made the paper were usually connected with these celebrations. The Battle Mountain *Scout* announced that, on

Sam Seals rode his special Midas-style horse in the 1987 Fourth of July parade. (Photo courtesy of Friends of Midas)

Kids' games and races are still popular in the 21st century (Photo courtesy of Friends of Midas)

Ms. Midas 1986, Audrey Fitzgerald, is pictured with attendants Ethyl Haltom and Ida Unger. (Photo courtesy of Friends of Midas)

Midas matriarch, Edna Wilkerson Timmons, occupied the Ms. Midas throne in 1987. Attending her on the float were (from left to right) Gina Laucirica, Missy Scheuller, and Jenny Scheuller. Mandy Scheuller is just out of view. (Photo courtesy of Friends of Midas)

July 4, 1927, the Midas team was whipped by the Betty O'Neal team, 22 to 2. Two years later, Midas avenged itself, beating Betty O'Neal at the Fourth of July game, 9 to 6. On another day in 1929, the Midas team clobbered Battle Mountain, 24 to 9.

Bob and Ida Unger helped to revitalize the Independence Day celebrations in Midas when they owned the Midas Bar in the early 1970s. Horseshoes replaced baseball as the main tournament, and children's games became popular. In the 1980s, a parade was added to the celebration. For several years, beginning in 1985, the parade was presided over by Ms. Midas; some of the women who have occupied this popular float were Joan Bennett, Audrey Fizgerald, Ethyl Haltom, Rosalie Rowe, Paula Scheuller, Edythe Seals, Edna Timmons, Ida Unger, and Bev White.

As the celebration became more popular in the late 1990s, volunteers have chaired the event. These have included Jim and Mary Schennum (1996-1997) and co-chairs Rich and Tawny Bennett and Kris and Debbie Mickelson (1998). Since 2000, Pam Mortara has spearheaded the festivities, which were held at the Midas School for five years. In 2000, Gene and Barbara Casci provided a barbecue grill from which the Friends of Midas group has served hot dogs and hamburgers. Brian and Theresa Elmore donated new horseshoes to the popular tournament. In 2005, a wiffle ball field was built next to the schoolhouse. That year, 175 people attended the day-long celebration.

Despite the loss of the schoolhouse, the 2006 Independence Day festivities were well-attended. George DeLong generously allowed the use of his garage to which everyone retired after the parade for the usual games of horseshoes and bingo. A new addition was the recognition of the best floats in the parade as determined by popular vote. First place was won by brother and sister Eric and Kalli Wegren with their humorous deer-hunting float; second place went to Julie Mazzone and Ben Tedore with their "Ben-Her" chariot. The fun-filled day was capped by an outdoor movie in Richard and Tawny Bennett's yard.

Church

Although the town had a social hall and a baseball field, no church was ever built in Midas nor has any report survived of a pastor attempting to establish a congregation. During the early years, visiting preachers traveled to Midas to lead the faithful in church services. These visits were usually limited to once a month, but weekly services may have been held during various periods. All of the visiting pastors whose names were recorded by "Midas News Notes" were from Winnemucca, including Reverend H. Lascella, Reverend Carl Fisher, Reverend A.R. Shivers, and Reverend Leo DeMare. Some Sundays, the mostly nondenominational church services were held in the Town Hall; other Sundays would find them in the schoolhouse. Occasionally, a denominational service was held. In 1931, the Episcopalian Bishop of Nevada, Reverend Thomas Jenkins, attended Midas services provided by Reverend Lascella.

Sunday school was also conducted in Midas. In 1933, "Midas News Notes" recorded that the Midas Sunday school had services and exercises on Memorial Day. Following the ceremony, the members of the Sunday school "decorated and fixed up all the graves." Early in the spring of 1936, the columnist reported that church and Sunday school attendance was good, despite the bad weather and announced that "our superintendents of Sunday School, Mrs. Grace Clawson and Mrs. Jacobson, are planning many outdoor recreation for the Sunday School as soon as the weather permits. Midas is fortunate in having such good teachers." Summertime would also find many of the town's children participating in vacation Bible school. Some summers the program, usually a week in length, was taught by Rev. Shivers; others were led by women who lived in Midas, like Miss K. Gordon and Miss Lota Hargrove who conducted Bible school in July 1938. Reverend Shivers also contributed to the social activities of the town by speaking about his visits to foreign lands, illustrated by pictures of these exotic places.

Frank, Hilda, and Victor Jacobson pose for a family portrait in their Midas home. Mrs. Jacobson was particularly fond of the piano that sits behind Mr. Jacobson. (Photo courtesy of Anna Jossis Hinrichs, Friends of Midas Collection)

This home was built for Mr. and Mrs. Noble Getchell in 1927 just north of where Skip and Dory Powell's house now stands. After the Getchells left the community in 1935, the house was occupied by Mr. Richards, Gold and Silver Circle Mine Superintendent, and later owned by Jack and Noreen Wilkerson Murdock. Dane and Wilma Foresight purchased the building and moved it to Bridge Street in Winnemucca where it is still used as a residence. (Photo courtesy of Lyon Family, Friends of Midas Collection)

Recreation

In the 1950s, Nevada travelers became more interested in the state's ghost towns, and articles describing these places appeared frequently in magazines, newspapers, and travel books. Midas was often included in these accounts. In subsequent years, numerous reporters visited to the town to research and retell its story. Over the past 50 years, the town has been featured many times in various publications, including most of the state's major newspapers.

The media attention attracted many visitors to Midas, and the area boomed as a popular recreational destination. Some people came as part of an organized photographic expedition; others simply loaded up the four-wheel drive to explore the area. As a result, Midas has become the base for numerous recreational activities, including motorcycling, snowmobiling, mountain biking, hiking, rockhounding, hunting, and fishing.

Although some of these activities are new to the area, the region around Midas has long been known for good hunting and fishing. Visitors to Noble Getchell's home in the 1930s came primarily to hunt. In the 1950s and 1960s, hunters rented cabins from Midas residents, such as Martin and Annie Brown, some of whom had come to town with the first mining boom and stayed after the mines closed. The game that interests most Midas hunters include deer, sage hen, and dove. Interestingly, one of the most sought-after birds is a relative newcomer to the region. Chukar was introduced to Humboldt County in 1935 and Elko County in 1938. The first hunting season in Elko County for this popular bird was in 1955.

Fishing has been a favorite sport in the area around Midas since the town's founding. On April 24, 1908, Elko's *Weekly Independent* quoted Sheriff Sweeney of Eureka in an article titled "Wonderful is Gold Circle." Among his many stories about the wealth of Gold Circle was a discussion of fishing. According to Sweeney, "it is the most wonderful fishing grounds in the world. You don't need any bait at all; you just cast the bare hook and line in the stream and you can catch a mess of trout in short order." Upon further questioning, the Sheriff claimed that the fish are "about as big as a horse." Although fish are plentiful in the area's streams and at nearby Willow Creek Reservoir, the small sizes of these bodies of water cause one to wonder whether Sweeney's story was just another fish tale.

As a center of hunting and fishing for many years, Midas is a logical place to hold an event to raise funds for wildlife preservation. The first weekend in August has been dedicated to this activity for many years. At first, Ducks Unlimited was the beneficiary of this popular event, which was chaired by Dick Rowe from 1998 through 2000. In 2003, the Midas chapter of Nevada Bighorns Unlimited (NBU) took over the dinner, games, raffles, and auction. Until 2006, Friends of Midas hosted the Sunday morning breakfast. It was not unusual for 150 people to participate in the dinner and auction on Saturday night and the hangover breakfast the next day.

Currently, the Board of Directors for the Midas Chapter of Nevada Bighorns Unlimited consists of Mark Barlow, Dan Bennett, Mike Bidart, Gene Casci, Gene Drakulich, Brian Elmore, Jeff Garejo, George Miller, Tony Scheuller, and Jim White. George DeLong serves as President of the board. The thousands of dollars raised on that weekend have gone far to help conserve habitat for Nevada's wildlife. Most recently, in January of 2006, the Nevada Department of Wildlife captured and relocated 30 Rocky Mountain bighorn sheep and 133 pronghorn antelope in northeastern Nevada. The Midas chapter of NBU provided a volunteer for the effort, Brian Elmore, and joined two other chapters to provide funding for the charter helicopter.

The Midas NBU group had its most successful event yet in 2006, raising 22 percent more than it did in 2005. Catered by John Lemich of Elko, the three-day event attracted over 200 people.

Many community events, such as this Ducks Unlimited dinner, take place at the DeLong garage. (Photo courtesy of Friends of Midas)

George Wingfield (on left) and Richard "Tamale Dick" Vanatta were business partners in Golconda around 1900. The bar on which they are leaning and the mirrored bar back behind them were moved from Golconda to Midas by Andy Jones in 1957 and may still be seen—and leaned upon—in the Midas Saloon and Dinner House. (Photo courtesy of the Nevada Historical Society)

Lt. Governor Denver S. Dickerson (Nevada State Archives)

Senator Key Pittman

Notable Visitors

The new district attracted the attention of mining men throughout Nevada, and some of the stampede to Midas included these people. One of Nevada's most prominent mining men, George Wingfield, was involved in the district's first year. The Elko paper reported in March 1908 that Wingfield and his partners bought the Gold Crown and Gift properties for $50,000 cash. Consisting of eight claims and 140 acres, both groups eventually became two of the biggest producers in Midas. Anecdotal evidence indicated that Wingfield visited the district on occasion, which is not surprising, considering his ownership of property in Golconda until 1910 and his close relationship with State Senator Noble Getchell who later was a prominent Midas resident. However, Wingfield's relationship with Gold Circle was casual at best: apparently, he sold the property rather quickly and was rarely connected with the district. Other well-known investors, named by the Tuscarora newspaper, included Tasker Oddie (elected Nevada's governor in 1910) and U.S. Senator George Nixon.

The Elko newspaper also chronicled the 1908 visit to Midas of Lt. Governor Denver S. Dickerson just two months before he became Acting Governor of Nevada upon the death of John Sparks. One of the men accompanying Dickerson was Key Pittman, "a well known attorney of Tonopah and Rawhide." It is possible that the voters of Midas favorably remembered Pittman's visit four years later when they joined the rest of the state in electing him to his first term as a United States Senator.

Nevada Governors

During the district's first one hundred years, three sitting Nevada Governors visited the district: James G. Scrugham (a Democrat who served from 1923 through 1927), Fred B. Balzar (Republican, 1927-1934), and Kenny C. Guinn (Republican, 1999-2006). Governor Scrugham visited Midas in May of 1924, according to the Battle Mountain *Scout*. While he was there, it was determined that he was the first governor to travel to Midas during his term of office. The paper observed that "other governors were there, but always either after or before their terms." To mark the occasion, Governor Scrugham was honored with a dinner and dance that continued until the "wee hours."

The next two gubernatorial visits coincided with mining celebrations. Governor Balzar participated in the dedication events for the Gold Circle Consolidated Mining Company in 1927; Governor Guinn helicoptered in for the dedication of the Ken Snyder Mine by Midas Joint Venture 72 years later.

To mark the initial success of the newly-organized Gold Circle Consolidated Mining Company and to celebrate the pouring of the company's first gold brick, Getchell sponsored a big party in Midas. According to the Battle Mountain *Scout*, on June 15, 1927, more than 400 people from throughout the state and the country traveled to Midas to join the celebration. At the ceremony, Governor Balzar poured the molten metal into the cast, "amidst the shouting and cheers of the visitors," to make the first brick. A millman poured the remaining gold to reach a total of four and a third bricks weighing 567 pounds and worth well over $20,000. The celebration was capped with a dance in the Midas Town Hall, which, the paper reported, was "gaily decorated for the occasion." Speeches were given by Governor Balzar, State Senator T. Tracy Fairchild of Elko County, and Getchell. The dance lasted until the early hours of the next morning.

When Governor Balzar died in office on March 21, 1934,

James G. Scrugham

Fred B. Balzar

Kenny C. Guinn

(Portraits courtesy of Nevada State Archives)

Governor Kenny Guinn (far right) enjoyed his visit to the beautifully restored Midas School during the dedication of the Ken Snyder Mine in 1998. Joining him were (left to right) Dan Bennett, President of Friends of Midas; Dana Bennett, FOM secretary; and Edna Wilkerson Timmons, the first student to graduate from the Midas School after it opened in 1927. (Photo courtesy of Friends of Midas)

the entire office staff of the Gold Circle Consolidated Mining Company attended his funeral in Reno. They were Getchell, Manager; Fred W. Varney, Mine Superintendent; Victor Jacobson, Mill Superintendent; and Charles Lyon, Auditor.

On Saturday, June 19, 1999, Governor Kenny Guinn and First Lady Dema Guinn joined 300 invited guests, including State Senator Dean Rhoads of Elko County, to dedicate Midas Joint Venture's Ken Snyder Mine. After arriving by helicopter, the Governor and First Lady toured the two-room Midas School that had been painstakingly restored by the mining company. There, the Governor spoke with Edna Wilkerson Timmons, the only person who had attended both mine celebrations and met both Governors.

At the Ken Snyder Mine site, the company conducted tours of the complex, including underground trips, and hosted a barbecue lunch. The program included remarks by Governor Guinn; Seymour Schulich, Chairman of Franco-Nevada Mining Corporation; Craig Haase, the company's Vice Chairman and General Counsel; and Andre Douchane, Vice President of Operations. *The Humboldt Sun* reported that Schulich had explained that "the Ken Snyder Mine was the one tangible result of a $25 million, 16-year exploration program that evaluated about 100 prospects." He indicated that construction of the mine created 390 direct jobs and an estimated 1,560 indirect jobs. At the end of his remarks, parts of which were cheered loudly by the Governor, Schulich presented Governor Guinn with a gold-plated hard hat.

The Manassa Mauler in Midas

Because of his political and mining contacts, Getchell attracted other notable people to Midas. Some visitors to the Getchell home in Midas traveled to the area to hunt sage hen. In 1931, hunters included Nevada Governor Balzar, California Governor James "Sunny Jim" Rolph, and Nevada Prison Warden Matt Penrose. But by far the most famous person to visit Midas and the home of Noble and Louise Getchell was former heavyweight boxing champion Jack Dempsey. In fact, the others visited Midas in order to hunt with Dempsey in October 1931.

Residents remembered Dempsey visiting Midas frequently during 1931. Several people recalled Dempsey jogging through town and using his greyhound dogs to hunt coyotes at nearby Squaw Valley Ranch. Byron Wilkerson recounted that Dempsey liked to stop in at Primeaux's store. Once, however, the visit was marred by an argument between Dempsey and Arthur Primeaux over the price of gasoline. Dempsey thought he was being overcharged for gas for his big Cadillac.

Despite this altercation, most people mentioned that Dempsey was a pleasant, friendly man who concentrated on his training: Getchell had set up a sparring ring for Dempsey near the Gold Circle Consolidated office. He also seemed to take a real interest in the young people of Midas. A favorite memory of Gordon Warren's was the "Dempsey Dance" held at the Town Hall in the champ's honor. The Battle Mountain *Scout* reported that the dance, held in the fall of 1931, was "said to have been one of the most pleasant recorded in that district."

Edna Wilkerson Timmons was hired to cook for Getchell's guests and to take care of the rooms when Dempsey was at "the big house." (Getchell's home, built in 1927 and palatial by early Midas standards, is now a simple home on Bridge Street in Winnemucca.) Timmons remembered that Dempsey "was always a gentleman. Very nice, very entertaining. And

Trips to Midas often provided photo opportunities for the Champ. The captions below and on the top of page 138 were originally written in 1931 for these International Newsreel Photos.

Be careful, Jack! You might blast your comeback plans!

Midas, Nevada .... Senator Noble Getchell (left) handing Jack Dempsey, former heavyweight champion of the world, some sticks of dynamite as Jerry the Greek, Jack's trainer, looks on. Jack has returned to the scene of his earlier labors in order to better prepare himself for the rigors of another climb toward the heavyweight title. He plans on spending at least 30 days at the Champion Mine, in which he is interested, before returning to the ring wars. (Courtesy of Friends of Midas)

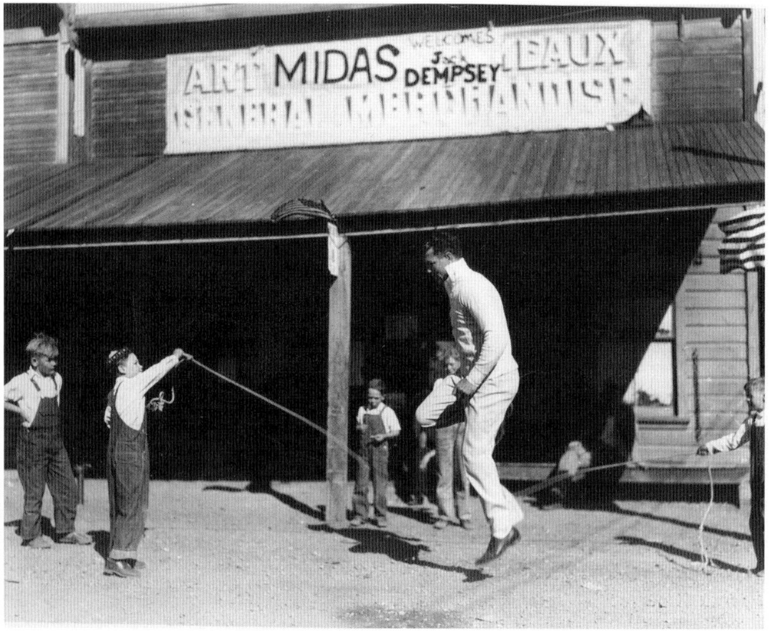

Jack makes Friends with Tiny Boosters in Mining Town

Midas, Nevada -- Jack Dempsey has returned to the mines again! This time to whip himself into shape for his current assault on the world's heavyweight championship which he once held. Jack is shown above skipping rope with some small citizens of this town to whom he is little short of an idol.
(The boy directly behind Jack Dempsey with curly hair is Loring Berg.)
(Photo courtesy of Friends of Midas)

This 1931 hunting party consisted of (left to right) Governor James Rolph of California, Governor Fred B. Balzar of Nevada, Nevada State Senator Noble H. Getchell, Nevada prison warden Matt Penrose, and Jack Dempsey. (Photo courtesy of the Lyon Family, Friends of Midas Collection.)

he was a teaser ... he'd help with the dishes." When Dempsey and the other hunting guests discovered that she would soon be leaving the area to go to school, they gave her a $20 tip. Considering that the average daily wage for Midas miners was under five dollars, this was a tremendous amount of money.

During the summer of 1931, Getchell joined Dempsey, Jim McKay (one of Reno's more notorious gambling men), and Tex Hall to lease a mine from Frank Macy. Called the Champion, the mine was to be a publicity vehicle for Dempsey and a profitable venture for all of them. In a letter to McKay on August 27, 1931, Getchell discussed the mine's progress. The tunnel was in 60 feet, two buildings had been built, and the blacksmith shop was near completion. Getchell noted that Dempsey planned to return to Midas after Labor Day to "muck in the tunnel for two or three weeks to strengthen his back and arms." Getchell also indicated that each of the Champion partners owed $562.50 in costs associated with opening and running the mine.

On October 14, 1931, the *Reno Evening Gazette*, in reporting about a house that Dempsey bought in Reno, noted that he had recently returned "from Midas where he spent a month working in a mine." However, it appears from Getchell's papers that Dempsey may not have paid his share of the operation's costs. By the time the mine struck some promising-looking ore in 1932, Getchell's correspondence did not mention Dempsey at all. Dempsey's lack of participation did not close the mine, however. On August 26, 1932, Getchell wrote to McKay and Hall (but apparently not Dempsey) about the Champion strike, noting that "the whole town here got drunk a few days ago in celebrating the ore body we have already opened up." Interestingly, this little party occurred over a year before Prohibition ended.

In 1931, Jack Dempsey enjoyed sage hen hunting during a visit to Midas. This picture may have been taken on the Squaw Valley Ranch a couple miles from town. (Photo courtesy of the Lyon Family, Friends of Midas Collection.)

This wintry view looks north on Main Street to Midas Mountain in the 1930s. (Photo courtesy of Edna Wilkerson Timmons, Friends of Midas Collection)

FIRE & ICE

Surprisingly, Midas has never been destroyed by fire. Countless other mining and ghost towns in Nevada were burned to the ground, but Midas, as a whole, has survived the various blazes that ignited and consumed individual buildings within the town. In the past 100 years, several fires and a devastating blizzard were recorded.

The first of the recorded in-town fires occurred on August 14, 1923. According to *The Humboldt Sun*, Vivian Cozart, 13 years old, had arrived in Midas a few days previously to visit her father, recently employed at the Red Hill Florence Mines. In trying to restart the cook fire to prepare lunch, Cozart apparently caused an explosion by throwing kerosene on coals that were, unbeknownst to her, still hot. The paper provides a detailed and grisly accounting of the scene, which resulted in the death of Cozart; severe injuries to L. Proebstel, a Red Hill miner who tried to rescue her; and the destruction of the old Halsell house (in which the Cozarts were living) and three cars. Vivian Cozart was buried in Los Angeles next to her mother who had died just a few months previously.

Two other houses were burned on January 2, 1930. At about 9:00 that evening, reported the Battle Mountain *Scout*, "Midas was aroused by a fire which burned two houses in less than an hour, one of the houses belonging to A.A. Primeaux and the other to H.A. Thompson. The houses were completely destroyed."

January of 1930 was also a snowy month. Noble Getchell wrote about it to a friend in San Francisco:

> The roads are fierce, snow and ice all the way, and the mail coming out of Midas on snow shoes . . . seven hours to go five miles from the Eastern Star camp to Midas the snow being from two to three feet deep. It now looks as though some little time will elapse before we can get any freight into Midas.

Both of these 1936 shots were made into post cards. (Courtesy of Friends of Midas)

Fire destroyed the Rock House around 1980. A heater in the house blew up, sparking a fire that was fought by John Sabin, Bryon Wilkerson, and Tim Timmons to keep it from spreading through the canyon. (Photo courtesy of Bob Lukens, Friends of Midas Collection)

The next fire occurred on July 28, 1933. The Battle Mountain *Scout* reported that the "residents of Midas turned out in a body to fight . . . flames of unknown origin [that] consumed the old Cody saloon and the Robert Richardson place adjoining." The paper commented that the entire town would have been in jeopardy had not the residents fought the fire so effectively. A vacant lot between the buildings also helped control the damage. Losses were estimated at $10,000.

Apparently, the fire started in the unoccupied Cody saloon, "noted early day establishment." It spread quickly to the Richardson's "soft drink and restaurant establishment." The paper related that Mike Paisley owned the Cody saloon and Tamale Dick owned the Richardson building (the site of Tamale Dick's rage two years earlier) and that "both buildings housed elaborate bars brought to the camp in the early days of the mining excitement." Sixty years after the fire, in May 1993, Richardson's son Don visited Midas. According to him, the rusted safe that has stood under the large cottonwood tree on the corner of Fourth and Main for many years was in his father's bar when it burned.

On October 14, 1933, one of Charles Warren's houses burned to the ground. The *Scout* reported that the house was occupied by George Holbert and that the cause of the fire was unknown. "The structure was a complete loss," the paper noted.

During the winter preceding the 1933 fires, the "Midas News Notes" concentrated on weather-related news. Reportedly "half the town" was sick with the flu, and blizzards had closed the roads. One January storm alone brought almost two feet of snow. Soon after that report, the Battle Mountain *Scout* observed that the snow was five feet deep at the Elko Prince Mine, and automobiles could not travel the roads. Horse teams were needed to go anywhere in the district.

Then, on February 3, 1933, a *Scout* headline exclaimed: "Midas Isolated As Heaviest Blizzard Of Year Hits." The story indicated that mail could not get to or from Midas for at least four days because of snow drifting across the road. Apparently, the snow had settled to four feet, but high winds caused significant drifting, making the roads virtually impassible.

Two weeks later, the Midas column cited several incidents that illustrated the difficulties the people in the district were experiencing due to the snow and high winds. The superintendent of the Gold Circle Consolidated, F.W. Varney, required over 14 hours to travel the 63 miles between Midas and Winnemucca, covering the first 15 miles from Midas on horseback. Gordon Warren and J.R. Purdy went to work on skis. The Wilkersons found that it took all day to complete a trip of five or six miles to haul hay from Squaw Valley for their milk cows. The columnist explained that people had to walk to get anywhere, and one man "was stricken snow blind."

It appeared that the district would have other troubles as well. In early February, the *Scout* referenced a letter to Getchell in which the snowy conditions were blamed for his employees having to walk to work. Some of the men were attacked by hungry coyotes, Getchell's correspondent continued. The men refused to walk alone, so they walked in groups armed with clubs and guns to protect themselves. The article noted that the reporter could not confirm this "coyote yarn."

The reporter was right to be suspicious of the story. On February 24, 1933, the *Scout* informed readers that the incident was "explained now by information of a much less alarming nature." The article continued:

> The big round, glowing eyes that peered from the snow upon the miners as they went to their chores and sent cold chills up and down the workers' backs, have turned out to be none other than the eyes of "Rusty," H.A. Thompson's big yellow house cat.

The newspaper had traced the story to Reno "where Senator N.H. Getchell had something to do with it."

Although the residents of Midas were not, apparently, facing hungry coyotes, they were concerned about other problems. The storms had delayed the delivery of coal to the

The first floor of the Gold Circle Consolidated Mining Company's office can barely be seen behind the snow. (Photo courtesy of Don Richardson, Friends of Midas Collection)

Noreen Wilkerson and Theda Phillips are ready for a fun day of sledding early in the 1930s. The Midas Café is in the background. (Photo courtesy of Edna Wilkerson Timmons, Friends of Midas Collection)

town, causing the price to escalate to as much as $40 per ton. By the end of February, the Battle Mountain paper stated that Midas had a week's supply of coal left, unless conditions changed. On March 22, the townspeople took matters into their own hands--literally--and "a large crowd of men shoveled out the snow to open [the Midas road] to cars." This huge chore took three days to complete. In some places on the road, the snow was four feet deep. The Midas columnist reported that "several of the men became snow blind while others received severely blistered faces."

Soon after this event, the "Midas News Notes" began recording the arrivals and departures of several people, indicating that conditions had, indeed, improved. However, the columnist also noted that a ranch foreman, traveling with eight horses, needed three days to complete the 30 mile trip from the IL Ranch to Midas. But the worst of the winter weather was over, and normal travel to and from the district resumed.

The next in-town fire was reported three years later. According to the "Midas News Notes," the Wilson's unoccupied house and butcher shop burned on November 28, 1936. The column did not indicate if the town was threatened in any way by this particular fire.

In February 1937, the *Reno Evening Gazette* reported that an excessive amount of snow had kept the three True trucks from getting into the town for two weeks. Eight-foot drifts had closed the road approximately eight miles from the town of over 200 residents, and the trucks were stuck in snowdrifts. Equipment from the State was required to open the road on February 14 after the county equipment proved inadequate. The newspaper explained that:

> One of the difficulties which has beset communication the last week with Midas is the fact that the telephone operator there cannot ring Winnemucca, but must await a call from here in order to relay messages from the snowbound camp.

More snow was expected within the week.

Two years later, another natural disaster loomed for the district. As were many northern Nevada towns during the 1930s, Midas was threatened with an infestation of Mormon crickets. On May 19, 1938, the "Midas News Notes" reported that cricket crews were stationed near the Clover Valley ranch not far from the town. The columnist remarked that "the crickets are just a short distance from here and more plentiful than last year," but did not indicate if the town ever suffered from the insects. In the summer of 2005, however, the crickets did overrun the town and surrounding hills.

In later years, range fires came perilously close to destroying the town. On July 24, 1942, a headline in the Humboldt *Star* exclaimed that "Midas Is Safe After Flames Threaten." A huge range fire had burned over 20 miles of land north of Midas and had come within four miles of the town. It burned in Squaw Valley up to the fence of the ranch and came close to the Esmeralda mine east of town. At one point, the prospects looked so grim that the residents were packed and ready to leave at a moment's notice. Exhausted firefighters could not keep the fire from jumping the road, but a passing Bell Telephone crew stopped to help fight the blaze. The newspaper suggested that the town might have been lost if the crew had not assisted.

Less than a month later, the Humboldt *Star* reported another fire in the area. Apparently, the assay office at the Esmeralda exploded early in the afternoon of August 1 and set off a 25-mile range fire. No one was hurt in the explosion, but it brought the fire within one mile of Midas. One hundred men fought the fire. They dug a trench around Midas with a bulldozer and started backfires to protect the town. After a three-day battle, the fire was brought under control, and the town was saved.

Snow could still cause concern in a Midas winter. The *Nevada State Journal* reported on the first day of 1946 that "the Midas community [was] completely snowbound last week for

Beginning in the late 1970s, snowmobile tracks were often seen in the road that skirts Midas Mountain. (Photo courtesy of Friends of Midas)

Tom Turk with the Nevada Division of Forestry makes a presentation to Gene Casci (left) in front of the Ron Hasson Midas Fire Station, which stands on the spot previously occupied by the Midas Town Hall. (Photo courtesy of Friends of Midas)

four consecutive days" Drifts up to three feet high had closed the road from Winnemucca. The article noted that "a mail truck, loaded down with Christmas gifts and toys, was stalled three miles from Midas" and concluded:

> Telephonic communications remain open with Midas and snow-bound adults, wanting to enjoy a Christmas themselves, said that efforts will be made to bring the holiday gifts into town by sleigh. Snow was still continuing to pile up there.

A range fire was finally controlled on Labor Day, 1950, after burning two and a half miles down the canyon south of the town of Midas. For a brief time, the fire had threatened the town, but the *Nevada State Journal* indicated that only range and telephone poles had been destroyed.

In October 1955, a fire started by careless deer hunters burned 500 acres in Midas Canyon. Within 10 days, a 1,200-acre fire in the upper Clover Valley was also being blamed on hunters. The hunters were not identified in news articles, which noted that they were being sought by federal investigators for prosecution. Local ranchers threatened to close the area to hunting until the arrival of a storm could reduce the fire hazard.

Evelyn Sabin recalled two subsequent fires. One destroyed several buildings, including some of the nicer homes in Midas, just a few days before she and her husband first arrived in Midas in 1956.

The second was a more spectacular blaze. Late in the fall of 1964, the school's teacher, Janice Benninghove, left for the weekend, leaving just a handful of people in town. According to the Elko *Daily Free Press*, at about 9 p.m. on Saturday, November 14, a fire of unknown origin began in her rented house, which was on Main Street just north of the old Warren Mercantile. By the time the fire was spotted, the wind, blowing fiercely from the north that night, had pushed the flames to the buildings south of the house, including the Mercantile and the Massey Building, an old warehouse on the south side of the store.

The warehouse stored some old mining equipment, including eight cases of dynamite. When the dynamite exploded, a handleless, iron pick flew from the warehouse and impaled itself in the ground just in front of John Sabin, Andy Jones, and Bill Hartzler, who were standing across the street by Jones's bar, watching the flames. The windows in the Midas Bar and other buildings around town were blown out, and whiskey bottles were destroyed. Annie Jones was hit in the face with the shattering glass and required subsequent medical attention in Elko. Adding to the fireworks, a propane tank fell over and was ignited by the blaze. The tank took off like a rocket and went flaming into the night sky.

The paper reported that all of the "buildings within an 800-foot radius were destroyed or extensively damaged." Despite the strong winds, the entire town did not go up in flames because several inches of snow had fallen the day before the fire. The next day the flying tank was found several miles away, galvanized iron pieces were seen wrapped around tree limbs like aluminum foil, and the wayward pick could not be pulled from the ground.

Raging range fires burned close to Midas again in 2001, 2005, and 2006, which was the worst fire season in Elko County history. The largest, the Winters Fire, burned over 240,000 acres within a perimeter of 174 miles and came within three miles of Midas, but the tremendous efforts of the firefighters kept the town from danger. Although the residents have been affected by heavy smoke and falling ash, no fire has yet required the evacuation of Midas. It may be argued that the most devastating fire in recent years was not any of these range fires, but the one that destroyed the historic and much-loved Midas School on November 6, 2005.

Midas existed for 84 years before a fire station was built. In 1991, the Nevada Division of Forestry constructed a wildland fire station in Midas. The building stands on the location of the Town Hall, which was torn down in 1986 for safety reasons. Ron Hasson, a Midas resident for many years, was instrumental in establishing the station, which was named for him. The building is owned by the Northeastern Nevada Fire District and was staffed and equipped during fire season by the U.S. Bureau of Land Management. Although the presence of the firefighters and their equipment was particularly reassuring to Midas residents during the hot, dry summer months, the station served all of northern Nevada until the new BLM station was built about two miles from Midas. The building has also been used as a community gathering place for water cooperative meetings, holiday celebrations, Elko County Commission hearings, and other activities.

In earlier years, before the station was built and fire hydrants were installed, the firefighting capabilities of Midas were haphazard at best. No organized firefighting force existed. Former residents recalled the existence of a small handcart that was utilized to fight the occasional fire. Sometimes a loosely-organized bucket brigade was also called into service. Often, however, the only recourse was to let the fire burn until it was extinguished naturally. Again, it is amazing that the numerous fires that occurred throughout the history of Midas consumed only a few buildings each time and not the entire town.

In 1999, ten Midas property owners gathered to organize the Midas Volunteer Fire Department. The first elected officers were Dick Rowe, President; Pam Mortara, Vice-President; and Jacque Ewing-Taylor, Secretary/Treasurer. The group solicited membership among the remaining property owners, noting that members would be provided firefighting training from the Nevada Division of Forestry. As of September 2005, the officers are Jerry Singleton, President; Lynn Isaeff, Secretary; Dwayne Jackson, Fire Chief; Gene Casci, Assistant Fire Chief; Pam Mortara, Trustee; and Jacque Ewing-Taylor, Trustee. The Midas Volunteer Fire Department maintains the Midas Fire House and firefighting equipment, which includes two small tanker trucks.

After the last inmate was released, the Midas Jail was left open to the elements. (Photo courtesy of Friends of Midas)

MURDER & MAYHEM

Criminal activities came early to the area. In February 1908, the *Elko Weekly Independent* stated that "Gold Circle is rapidly making a criminal record that bids fair to equal that of any mining camp in the state."

The first crime reported by the newspapers was a duel between Harry Stuart and Bob Hart on Monday, February 10, 1908. Bell Macy noted that the shooting took place in Summit. Apparently, Stuart was the original locator of the Cleopatra claims, but Hart believed that Stuart had not done enough work to hold them. So Hart relocated the claims and was working them when Stuart approached him. A duel ensued: both men claimed the other shot first, and both were wounded.

When Elko Sheriff Clark returned to Elko from Golconda where both men were under medical care, the *Elko Weekly Independent* printed: "He reports Hart and Stuart, who shot each other at Gold Circle, getting along nicely."

The Battle Mountain paper commented that "this is the first shooting scrape in the camp, although there have been frequent gun plays." The event resulted in the organization of the committee to arbitrate claim disputes.

The First Murder

Murder was first committed in Midas two weeks later. On February 26, 1908, one of the saloons was the scene of an argument about money between Thomas R. Bean, a 42-year-old miner from Colorado, and George R. Dove, a 34-year-old teamster from Tuscarora. Eventually, the men left the bar. Later that night, Dove pawned Bean's watch at another saloon and gambled with the money.

Their suspicions raised by Dove's actions, concerned citizens soon found Bean's beaten body lying in the snow barely 40 feet from the first saloon. His face had frozen to the ground, and hot water was needed to release it. The murder weapon, a club covered with blood and hair, was found near the body. Dove was quickly arrested and just as quickly taken to Golconda as "threats of lynching Dove were freely indulged in."

At first, Dove denied killing Bean. Soon, however, he confessed to beating him to death with a "willow pole" and entered a plea of guilty to a charge of first-degree murder. Although he later rescinded his confession and claimed to have acted in self-defense because Bean drew his revolver, Dove was sentenced in April 1908 to life imprisonment at the state penitentiary in Carson City. Bean had been in Midas only two weeks. His body was sent to his wife and children who had remained in Colorado.

As newspapers use sensational stories today, the Elko *Daily Free Press* used the Bean murder to sell papers. At one point, the Elko paper asked its readers: "Did you notice that the Free Press had the first and only detailed report of the Gold Circle murder?"

The *Elko Weekly Independent* did not take such a sanguine view. On March 6, 1908, the paper expressed the following opinion, which the writer appeared to hold strongly despite questionable abilities at arithmetic:

> It is reported that five saloons are doing business at Gold Circle. The population of the district is estimated to be between 250 and 300. Taking the larger number as the correct figure there is a saloon to every 6 persons in camp. If all these saloons run a year, the county will receive $600 from them in licenses, not enough to pay one quarter of the expense of one criminal trial, caused by the excessive use of whiskey or other intoxicants. There should be no more saloon licenses issued for Gold Circle or not at least until the population increases many fold. More saloons make more misery and crime and more expenses for the taxpayer to pay.

Dove did not spend the remainder of his years in the state pen. He was paroled on May 16, 1917, into the custody of Messrs. Kirman and Lewis to be employed on their Smith Valley ranch. A few months later, he was allowed by the Board of Parole Commissioners to accept employment with anyone in Washoe, Ormsby, Douglas, or Lyon Counties. He was required to keep in touch with the board and to stay out of Elko County.

His requests for parole began in 1912, accompanied by a petition signed by 40 people. Several letters also were sent on his behalf, including one from his brother, renowned Nevada mining engineer Thomas Wellington Dove of Tuscarora, who stated that "liquor is his only downfall." George Dove's supporters claimed that, if he had not been under the influence of alcohol at the time of the crime, he would not have killed Bean. They pointed out that he had not been involved in any prior criminal activity and had exhibited exemplary behavior while in prison. Dove's requests for parole promised that he would abstain from "intoxicating liquor." After being paroled, Dove usually included in his letters to the board a note confirming his continued sobriety.

On November 15, 1921, Dove received a full pardon and the restoration of the rights of citizenship, including suffrage. The pardon was granted, in part, because Judge E. L. Taber of Elko sent a letter on Dove's behalf, noting the "blighting influence" the crime had on Dove's two sons. Judge Taber requested the pardon for the boys' sake; he had met them and "found them to be fine appearing and gentlemanly young men."

This crowd is preparing to remove two Chinese men from the Midas Jail and force them to return to Tuscarora. People of Asian origin were prevented from settling in Midas; consequently, unlike many other mining towns (such as Tuscarora), Midas did not have a "Chinatown."
(Photo courtesy of Edna Wilkerson Timmons, Friends of Midas Collection)

Building the Jail

The Elko County Commission responded to the criminal activities early in 1908 by directing the Sheriff to appoint a deputy sheriff for Midas who would be paid $100 a month. Additionally, the commission authorized the building in Midas of "a small jail, suitable for present needs." On March 3, 1908, Guy Harbin was appointed deputy sheriff. The *Daily Free Press* stated that "Harbin is a good man for the place." One of Harbin's first duties was to complete the jail. Apparently, he went straight to work on the building: the Midas report to Elko's *Daily Free Press* observed on March 18, 1908, that "our Deputy Sheriff Guy Harbin is busily engaged building a town bastile." The one-room structure was uniquely constructed by placing two-by-four boards on their sides and interlacing them at the corners. By the county commission's May meeting, $437.02 had been expended in labor and supplies on the jail, which still stands.

The murder of Bean was not the last violent crime in Midas in 1908. On May 14, the Battle Mountain paper reported that Lee Ballard used an axe to attack an "old man" in Gold Circle. Ballard allegedly was trying to rob and murder the unnamed victim, "but the latter wrenched [the weapon] from him and prevented his being able to carry out his purpose." Ballard quickly escaped Gold Circle, but was later arrested in Battle Mountain.

Almost three weeks later, on June 2, 1908, Deputy Sheriff Harbin arrested three men for highgrading, a serious crime in the eyes of mine owners. Highgrading was the process by which employees of a mine hid pieces of gold and ore in their clothes and lunchboxes when they left work at the end of their shifts. The activity led many mine owners to add a changing room where workers were obliged to change between street and work clothes at the beginning and end of each shift.

Apparently, the St. Paul Gold Mining Company had not yet installed such a room at their Midas operations. The unnamed men absconded with over $2,000 worth of gold from the company. These men were probably among the first visitors to the town's new "bastile."

In spite of these events, criminal activity was not common in early Midas. In the June 20, 1908, issue of *The Gold Circle News*, these entries were included in the "Gold Circle Nuggets" column:

> Quietness reigns around Justice E.C. Rickard's quarters. The judge laments it is distressingly quiet in the matter of "justice dispensation."
>
> We asked Guy Harbin the other day: any new comers at the fort? And Guy grimly answered "Nay, I am holding the fort all by myself with no deeds in the camp that requires [sic] application for admission." And so the jail is still empty of enforced boarders.

Subsequent years were fairly quiet. No additional criminal shootings were reported until 1931, but the town was home to a thief and a draft-dodger.

On August 5, 1917, an Elko *Independent* headline exclaimed, "Midas Company Robbed Of Big Amount Of Gold." The article reported the recent discovery at the Elko Prince Mill that "the amalgam plates that catch the gold from the slit in the big mill were being systematically pilfered." Apparently, the owners had suspected that the activity was going on, but had delayed calling the authorities because they thought they could catch the guilty employee. The paper noted that the thief "knew his business" and was not caught.

The Elko County Sheriff and District Attorney traveled to Midas immediately and planned to stay until the case was solved. They estimated that the company probably lost $5,000. The report stated that "no names of suspects have yet been divulged by the officers but it is believed that evidence pointing to the guilt of certain prominent persons in Midas is strong." Unfortunately, the newspaper did not print any subsequent articles.

On May 23, 1919, the Elko *Daily Free Press* reported that a Midas resident was arrested for failing to register for the draft. Although World War I had officially ended six months earlier, Malcolm McLeod found himself in the county jail for being a draft dodger.

Murders of Young and Johnson

The summer of 1931 brought two brutal murders into Midas. Considering the small number of people who lived there and the infrequency of murder throughout its history, these bloody episodes must have rocked the canyon. The first murder victim of 1931 was Dell Young, 45. He was shot in the back of the head shortly after 6 p.m. on Saturday, May 23, 1931, with a .38 caliber revolver. Although the weapon was Young's, his assailant was unknown.

According to the Battle Mountain *Scout*, Young, his wife Mary, and his brother Jack had moved to Midas early in 1930 from southern Nevada to work a lease on a claim. The lease expired on May 1, 1931, and Dell Young had not been working. He and his wife were seen in the main part of town approximately two hours before he died, walking to their cabin, which was located about a mile down the canyon from the main part of town. He had been drinking heavily, although the town was officially "dry" due to Prohibition.

Mary Young later claimed that she was working in the garden when she heard the shot. She ran inside and found her husband dead. She picked up the gun lying next to him, then dropped it. It landed in a pool of blood, which splattered her dress. She brought Constable Paisley back to the house, telling him her husband had committed suicide. However, on June 13, 1931, based on the findings of a criminologist, Young was arrested and charged with murdering her husband.

She was tried in Elko that fall. One of the noteworthy aspects of her trial was an admonishment from the judge concerning the county's lack of facilities for a jury consisting of men and women. On November 6, 1931, the Battle Mountain *Scout* reprinted the remarks of Judge L.O. Hawkins of Winnemucca when two women were seated on the jury. At that time, the judge scolded Elko County for being unable to sequester a jury that included women "without violating the rules of common decency." He pronounced that "for 10 or 12 years it has been the law of this state that women are competent to sit as jurors in all cases." Hawkins directed the county to remedy the situation immediately, although the article did not specify the problems that precipitated the judge's reprimand.

Several Midas townspeople were called to testify, but the evidence against Young was entirely circumstantial and complicated by a set of bloody footprints that were seen going from the back of the house and into a gully where they disappeared. The jury deliberated for four hours and acquitted her on November 8, 1931. No one else was charged with committing this crime. Apparently, Young continued to live in Midas for some time after this incident.

The second murder occurred around 3 a.m. on August 5, 1931. Sanfrid Johnson, 29 years old, was killed instantly by a shotgun blast to the face on Main Street in Midas. The gun was fired by Richard "Tamale Dick" Vanatta, 69 years old. Vanatta had been in the area for many years; in fact, he was George Wingfield's partner in several Golconda race horses in 1899.

The Battle Mountain *Scout* reported that Vanatta was gunning for Paul Webber, manager of the Midas pool hall, who was Vanatta's "alleged competitor . . . for the Midas bootleg trade." The paper later stated that the two men were competing over the pool hall and soft drink business. Vanatta, angry with Webber, vowed to kill him. Vanatta got his shotgun and fired at

Webber in the street, hitting him in the back. Webber ran into "an establishment" where Johnson was playing cards. The next shot hit Victor Berg in the arm. The third shot caught Johnson as he stepped into the street.

The report from the Elko County District Attorney indicated that the incident was provoked by competition between the two men over the soft drink trade in Midas because they operated adjoining establishments on Main Street. Considering the ferocity of this conflict, which occurred during Prohibition, it is difficult to believe that the two men were battling over the sale of cola. Most likely, the allegation that they both sold illegal alcoholic beverages was accurate.

The District Attorney's report (in which the names of the participants are spelled differently) stated that, indeed, alcohol was involved in the events of that night. The document also provided additional details, which are reprinted here:

> On the night of August 4th there was a dance in Midas and after the dance in the early morning of August 5th the orchestra and several other people went to Paul Weber's place of business. The defendant who was quite drunk and from the evidence we gathered had been continuously drunk for three years went in and accused Paul Weber of stealing customers from him. There was a small argument in which Weber ordered defendant out of his place of business and when defendant refused to go hit him over the head with a pistol; then defendant went out and procured a shotgun and returned to the back of Weber's place of business, and when Weber was standing in front of the bar with his back towards him fired the shotgun at him and peppered his back with buckshot. Weber was not seriously hurt however. Victor Berg who was in Weber's soft drink parlor stood in the door and threw two beer bottles at the defendant and while in the act of throwing the third bottle defendant fired a shot again and tore off part of his fore-arm. The defendant then went to the front of Weber's soft drink parlor where one or two parties tried to take the gun away from him; and he advised them to leave him alone or he would give them some also. At this time Sanfrid Johnson, the deceased, walked around the corner of Weber's building and

defendant fired his gun. A full load of buckshot went into Sanfrid Johnson and he died instantly.

Vanatta had mistaken Johnson for Webber. In killing the Finland native who had been in Midas for two years, Vanatta killed one of his best friends.

Noting that Vanatta was "a cripple and had been constantly drunk for three years," District Attorney J.C. Clark reported that Vanatta was arrested on a charge of first-degree murder on August 19 and faced a preliminary examination in the Justice Court of the Midas Township on that date. On September 23, 1931, in Elko District Court, Vanatta pleaded guilty to a bargained charge of manslaughter and was sentenced to a term of five to ten years in the state penitentiary in Carson City. Justice E.P. Carville stated at the sentencing "that he had investigated the case and that he would not dismiss the murder charge filed against Vanatta but would sentence him on the manslaughter basis, thus establishing a precedent in Nevada criminal procedure." Apparently, the judge considered several extenuating circumstances in this case, including Vanatta's relationship with the victim, the obvious lack of intention to harm Johnson, the lack of prior convictions, and Vanatta's good reputation during the many years he had resided in Golconda.

Tamale Dick arrived at the state prison on September 27, 1931, where he died from cancer less than a year later on August 26, 1932.

Sanfrid Johnson, whose Swedish surname was Eurs, was buried in Battle Mountain. He was the brother of Anna Eurs Berg, who was the wife of Oscar Berg and mother of Astrid, John, Leonard, and Loring. Victor Berg, who lost part of his forearm to Tamale Dick's shotgun blast, was Oscar's brother, their Swedish surname was Glasberg. Oscar had changed to Berg when he arrived in the United States in 1912. Victor did not change his surname, but used Berg only in Midas, according to family sources.

Sanfrid Johnson is wearing the light-colored coat. The other man is unknown. (Photo courtesy of Len Berg, Friends of Midas Collection)

Mayhem

Despite these heinous murders, Midas was not exactly a hotbed of illegal activities. The next reported crime occurred over four years after Tamale Dick's rampage. From 1935 through 1937, the town witnessed two burglaries and one jealousy-motivated shooting.

In September 1935, the town's lack of adequate law enforcement was illustrated by an article in the Battle Mountain *Scout* headlined "Midas Safe Yields $17 To Robbers." The article related that an unknown thief had blown open the safe in Primeaux's store sometime between 10 p.m. on Sunday, September 15, and 7 a.m. the following morning. Payday had been four days earlier, after which many people had paid their accounts at the store. Consequently, Primeaux held a large sum of money, but he had not placed it in the safe. No one claimed to have heard the explosion. The newspaper reported further that Sheriff Harris was called in from Elko as "no peace officer exists in Midas at the present time," but the *Scout* did not report the results of the investigation. Less than a month later, the county appointed Al Fox as the constable.

A year later, Midas was the location of another, rather unprofitable, burglary. In the summer of 1936, the "Midas News Notes" commented that, on June 30, "some unknown party broke a window in the Rossi saloon and gained entrance and took about $4 or $5 worth of nickels, besides some liquor." It is not known if the guilty pilferer was ever apprehended.

On December 30, 1937, the *Scout* announced the arrest of Joe "Hungry Joe" Horvath. He was booked at the Elko County jail by Sheriff C.A. Harper and charged with shooting Val Hunter, a Midas woman, through the leg. The paper explained the shooting:

According to the Elko officer's report, it is a triangle love

Hungry Joe Horvath. (Photo courtesy of Jack and Noreen Wilkerson Murdock, Friends of Midas Collection)

> affair in which a man by the name of Fred Erber figures. The story is that Horvath went to the girl's cabin to get Erber but he had left earlier and in the ensuing fight Horvath shot the girl in the leg, missing her with a second shot.

Observing that "Hungry Joe is a Midas character having mined in that vicinity for several years," the article stated that Hunter would not press charges against her assailant.

The Last Murder

The last murder in the canyon occurred in 1963. On November 22, the newspapers for the towns around Midas had two murders to report: the assassination of President John F. Kennedy in Dallas and the fatal shooting of Richard Wayne Robertson in Midas.

Robertson, 33 years old, had been a Midas resident for at least a couple of years. In August 1963, he allowed Mr. and Mrs. Alvin Burkett and their 14 children to move into a Midas house and did not require any rent payments. He also promised to employ Mr. Burkett, 48 years old, who was en route to Oregon for work. The relationship between the Burketts and Robertson, however, quickly deteriorated.

Nine days before the murder, Robertson forced his way into the Burketts' home while Mr. Burkett was working at a local ranch. Mrs. Burkett confronted Robertson with a shotgun, which he wrestled away from her. After the struggle, she was taken to Humboldt General Hospital in Winnemucca for treatment of a severe head injury (later diagnosed as a concussion) and a lacerated arm. She claimed that he beat her with the gun; he claimed that he acted in self-defense and that she was injured during the fight.

After Mr. Burkett returned home and learned of his wife's injuries, he fired a rifle at Robertson. Standing approximately 200 yards away, Robertson was not injured. Only one of the several bullets Burkett fired came close to its mark, passing

"Charlie," the permanent mannequin occupant of the Midas Jail, enjoys his luxurious accommodations. (Photos courtesy of Friends of Midas)

through the leg of Robertson's pants. Although Burkett traveled to Elko to file a formal complaint against Robertson, neither of the men ultimately pressed charges.

Robertson was known as a trouble-maker in Midas before problems arose with the Burketts. He often carried at least one weapon and was the subject of many complaints from residents to Elko County Sheriff Jess Harris, who traveled to Midas several times throughout 1963 to investigate. Robertson operated a bar in Midas until the county liquor board revoked his license after testimony about his actions. The board refused to approve his request to reinstate his license in time for the 1963 deer hunting season.

Early in the fall of that year, water service to Robertson's home was discontinued for failure to pay his water bills. He retaliated by using a deer rifle to shoot a hole in the main line of the town's water system. For this crime, he was sentenced to 30 days in the county jail. During his incarceration, he deliberately banged his head against his jail cell and engaged in other behavior that caused county officials to send him to the State Mental Hospital in Sparks for observation. He was diagnosed as "socio-psychopathic," but was inexplicably released after a treatment period of only two weeks. He returned to Midas a few days before his altercation with Mrs. Burkett.

At 8:30 in the morning of November 21, 1963, Robertson allegedly barged into the Burkett home without knocking and seemed to reach for a gun. Alvin Burkett stated that he warned Robertson to "hold it." But Robertson, who was, in fact, unarmed, "swung around as if he had a gun in his hand," so Burkett grabbed a pistol and shot him. Upon arriving in Midas, Sheriff Harris found Robertson with a .22 caliber bullet in his head and Burkett claiming self-defense. Burkett was held in Elko until a coroner's jury could be convened.

The inquest included testimony from Burkett, his son, Sheriff Harris, and Andy Jones. On November 26, 1963, the jury concluded that Robertson's previous actions caused

Burkett to act in self-defense and ruled that the shooting was justifiable homicide. In releasing Burkett, the Elko District Attorney observed that "Alvin Burkett could reasonably believe that he and his children were in danger of being harmed by Robertson." Robertson's body was sent to California for burial, and the Burketts soon left Midas.

Restoring the Jail

In the years since the Robertson murder, the Elko County Sheriff has been called to town to address various concerns, but none have resulted in serious bodily harm. The jail, however, has held only cobwebs and dust for the past half-century.

The building and the lot on which it stands was given to Friends of Midas in 2000 by Walter "Bud" Aldrich, who had owned them for about 40 years. Immediately upon acquisition, Midas residents began the restoration process. George DeLong used his equipment to clear the site of brush. Dick Rowe, Gene Casci, and Dan Bennett reconstructed the walls. Don Mellen and Dick Bade rebuilt the roof. Buck Tulley shingled the roof and oiled the entire structure. Don Mellen and granddaughter Anna Mellen attached a bunk to the wall and placed a jailbird—a mannequin dressed in period attire—on the bunk. Dick Nietfeld, a Nebraska blacksmith, replicated the iron hasp and lock. All of the labor was donated; materials for the restoration of the jail that originally cost $437 came to approximately $3,000. The tiny facility now looks much the same as it did 100 years ago.

Restoration of the Midas Jail was a community project in 2000. Dick Bade and Don Mellen work on the roof in the upper photo. (Photos courtesy of Friends of Midas)

Clearing the ground of heavy brush helped to locate the boundary of the original cemetery. (Photo courtesy of Friends of Midas)

Dan and Joan Bennett planted hundreds of daffodil bulbs that brighten the area every spring. The Gardiner G. Bliss marker is shown surrounded by the flowers. (Photo courtesy of Friends of Midas)

MIDAS CEMETERY

By April 1908, newspaper reports indicated that 1,100 people were in the new camp. The Midas correspondent to the Elko newspaper remarked: "Our death list numbers five and we have a population of 1,500. Health conditions generally are good, and this morning I don't know of a sick person in town."

Apparently, some concern existed about the conditions at the new town, compelling Henry Porch to comment about the issue to the Elko newspaper. Reprinted in the Battle Mountain paper on April 30, 1908, the article on Porch's visit to Midas consisted mostly of the following:

> Mr. Porch emphatically denied that unsanitary conditions exist at Gold Circle and the reports published to the effect that much sickness exists there he denounces as malicious lies.
>
> "Conditions at Gold Circle are more sanitary and healthful" he said "than in any camp in the west. We have the best pure mountain water in the state and the fellow who says that an unusual amount of sickness prevails is either a fool or a knave. We have had but two deaths in the camp and only one of these was from natural causes, the other as you know was the result of murder."

The records tell a different story from both of these accounts. Not counting the murder of Thomas Bean on February 26, 1908, six men had died in Midas by April 30, 1908. The number was tallied by Bell Macy in her day-by-day account of her first year in the new mining camp. She noted two more deaths in May. That first winter in Midas was clearly harsh as numerous men suffered – and many died – from pneumonia. At least two men, as recorded by Macy, were driven to commit suicide.

New markers memorialize those who have been identified as buried in now-forgotten plots in the cemetery. (Photo courtesy of Friends of Midas)

A cemetery is not mentioned in the Macy journal, but the Elko County death records identify six men who died and were buried in Midas in April and May of 1908. Bell Macy also did not list the names of the dead, so it is impossible to compare her list with the County's. Based on county records, it would appear that Edmund Deakin was the first occupant of the Midas Cemetery. However, he died on April 10; Macy noted that a man succumbed to pneumonia on April 7. This unknown man might be in the Midas Cemetery.

No one knows for certain how many people were buried in the cemetery in subsequent years. Apparently, no organized records of the Midas Cemetery were kept. Edna Timmons and others who saw the cemetery in the 1920s and 1930s estimated that 25 to 30 people may be buried there. Indeed, 30 people have been identified. Clearing the area of brush appeared to reveal 35 to 38 gravesites, which provides a probable maximum number of cemetery occupants.

The Midas Cemetery is located southeast of the main part of Midas. For most of the last decades of the 20th century, the cemetery was difficult to locate and inaccessible except on foot. During the town's heyday, however, it was a fenced, well-maintained area at the end of a short road. Most of the markers were made of wood and painted white; a few were marble. Former residents remarked upon the beauty of the place, and Timmons remembered that the cemetery could be seen when driving into Midas. Now, thanks to the efforts of many volunteers, the Midas Cemetery is, again, a well-maintained area at the end of a short road and the white fence can be seen from the road into Midas.

Death certificates issued by Elko County provide the most comprehensive source of identities for the occupants of the Midas Cemetery. Following are their names and backgrounds:

Edmund R. Deakin, 32 years old, who died on April 10, 1908, of pneumonia. He may have been the first person buried in the Midas Cemetery.

Arthur G. Neville, 35 years old, who died on April 12, 1908, of pneumonia. A married miner, Neville had been in the area 30 days, coming from Tuscarora. C.A. Babcock, a "registered embalmer," served as undertaker.

Alexander McBeth, 65 years old, who died on April 13, 1908, of pneumonia. McBeth, a widowed miner, had been in Midas just over three months. Babcock served as undertaker. On April 13, Bell Macy wrote in her journal that three men had died the previous night.

Henry P. Voss, 25 years old, who died on April 19, 1908, of pneumonia. He had also been in the area approximately one month.

Joseph La Breche, 45 years old, who died on May 1, 1908. No cause of death was recorded for the Canadian. C.A. Babcock served as undertaker.

Lorenzo Dix Pierce, 24 years old, who was buried May 24, 1908. No cause was recorded. On May 21, 1908, Macy mentioned in her day book that "another man committed suicide in Midas this morning." She does not provide a name; perhaps it was Pierce. Perhaps it was not. Pierce was newly arrived from Cherry Creek, Nevada. Babcock again served as undertaker.

Hobart B. Luther, 79 years old, who died on October 14, 1911. Not married, Luther was employed as a teamster and stage driver. He had been in Midas five months, having arrived to be cared for by his long-time friend Francis Coleman (grandmother to Timmons, Byron Wilkerson, and Noreen Murdock). The town's doctor, Edward Dunscomb, M.D., had cared for Luther over the month prior to his death. He succumbed to complications from diseases due to his advanced age and was buried in the cemetery on October 15, 1911. William Muffly acted as the undertaker.

The Humboldt Sun carried Luther's obituary on the front page under the headline "Pioneer Of '48 Dies At Midas." After a lengthy description of Luther's long life, the article, written by Fred J. Niess, described his funeral at the Coleman home where he had lived. At the service, Margaret Brady sang four songs and Charles E. Whelund delivered the eulogy. Friends and neighbors followed the body to the cemetery at which Brady sang "Lead, Kindly Light" and "Abide With Me." Taking a bit of poetic license, Neiss concluded Luther's obituary:

> On the banks of Summit Creek, near Midas, where the winds and the willows sigh a requiem and the streamlet sings a lullaby, kind friends gently laid away Mr. Luther to his eternal and peaceful rest.

(Photo courtesy of Friends of Midas)

Baby Warren, who was stillborn on April 5, 1912. The unnamed infant was the twin sister of Desda Warren and daughter of Charles and Ruth Birda Holbert Warren. She was buried the next day under the care of her great-uncle, Dr. Dunscomb.

Mary Louise West, 36 years old, who died on December 24, 1913. Her death certificate noted that she was a married housewife, but it did not name her husband. She had lived in Midas six months. The official cause of death was "abortion on or about December 4th." Because "abortion" is often the medical term used for "miscarriage," she may not have succumbed to an induced abortion, which was illegal in 1913. She had been under the care of Dr. Pettingill for six days before her death. She was buried on Christmas Day. No undertaker was named.

Edward Dunscomb, M.D., 77 years old, who died on June 6, 1915, from natural causes. The cause of death was certified by Frank Macy, Justice of the Peace. Under the direction of J.M. Wyatt, Dr. Dunscomb was buried in the Midas Cemetery on June 7, 1915. The Elko *Daily Free Press* noted that, as the only physician in town, Dr. Dunscomb "prescribed and took his own medicine" for his illness. He was not, however, able to render for himself another medical service, the authentication of death. The paper reported that the town's residents were "puzzled" about the proper recording of Dr. Dunscomb's death because he was the only one who could legally sign his death certificate. They consulted Elko's Judge Castle who advised them to hold a coroner's inquest. The paper continued,

(Photo courtesy of Friends of Midas)

"The deceased has lived in this country for a number of years and was an eccentric old fellow, and without a doubt his death was caused by the gradual breaking down of old age."

Born in New York in 1838, Edward Dunscomb graduated from New York University Medical College in 1863. Two years earlier, he had been honorably discharged as a Captain in the Harris Light Cavalry. Anecdotal evidence indicts that he provided medical services throughout the rest of the Civil War. He lived in Colorado from 1896 to 1905 when he arrived in Reno and was licensed to practice medicine. By 1907, he was in Tuscarora; a year later, he moved to Midas.

Dr. Dunscomb provided the necessary medical services to the residents of his new town by attending numerous births, illnesses, and some deaths. But he was also interested in mining activities in the district. In 1909, he and others attempted to establish the town of Dunscomb on the Lucky Chance claims, which he held about two miles from Midas. The town never took hold nor did he realize much from his mining claims.

Gardiner Geyman Bliss, 12 years old, who died July 11, 1915. He was the victim of acute rheumatic fever, according to Dr. Daniel Pettingill. He had lived in Midas three years and was buried on July 13. The undertaker was F.J. Benneson.

Artro Teniente, 28 years old, who was shot on June 10, 1916. Teniente was a Portuguese who was married and employed as a laborer. The verdict of the coroner's jury was that Teniente died from "the accidental discharge of a pistol in the hands of one Tony Teniente." He was buried in the cemetery the next day, Wyatt again serving as undertaker.

The day after the burial, the Elko newspaper reported the tragic accident. The dead man, a sheepherder, had been one of several men at a bonfire, examining an automatic revolver. The Midas correspondent to the paper explained that the accident took place at the head of Oregon Canyon and that the man responsible for pulling the trigger – a nine-year friend of the deceased – was "inconsolable." After noting that the Elko Justice of the Peace was conducting the inquest in Midas, the reporter solemnly concluded with a comment that still appears in accounts of accidental shootings: "Another case of not knowing it was loaded." Although the paper initially misidentified the man, the correct name appeared in the subsequent list of people who had died in Elko County in June 1916.

Emma Halsell, 49 years old, who committed suicide on November 21, 1917. The official cause of death was "mania." She was buried two days later with Frank Macy serving as the undertaker. According to the Coroner's Inquest, held in Midas on November 22, 1917, Mrs. Halsell had been suffering from "melancholia to mania which is very extreme destructive case of insanity" and had slit her own throat with her husband's razor. She died within minutes of the arrival of Dr. L.M. Seaton. Sadly, Mrs. Halsell had been discovered by her daughter.

Frank G. Olinn, 67 years old, who died on December 18, 1917, from pneumonia.

Mateo Francisco Azevada, 52 years old, who was buried on December 24, 1917, after a wagon accident. According to *The Humboldt Sun*, Azevada, a teamster, was driving a four-horse wagon from Red House to Gold Circle. Somehow, he fell and was run over by his own wagon four miles from Myer's Station. A six-year resident of Midas, Azevada was survived by his wife. His dog had been found waiting patiently by his body.

Baby Girl Morey, who was born prematurely on October 13, 1918, and died the next day. Her parents were Hugh and Genevieve Morey. Dr. Pettingill attended.

Cecilia Clawson Wagner, 19 years old, who died on December 23, 1918, three months after her birthday. The Elko *Daily Free Press* observed that she was the first Midas casualty of the Spanish flu epidemic that killed thousands of people throughout Nevada and the nation during the fall of 1918. Her funeral was held in Midas on December 25 and was most likely attended by the many members of the Clawson family who lived in Midas and Tuscarora during the early years of the twentieth century. Wagner was

These two headstones, shown in 1978, survived the years, probably because they were hidden in the brush. (Photos courtesy of Friends of Midas)

The Clawson family remembered their relatives with these beautifully carved wooden markers in the late 1980s. (Photos courtesy of Friends of Midas)

survived by her husband Milton and baby daughter Helen who was not quite a year old.

Carl W. Schein, 60 years old, who died on April 15, 1919. His death certificate, signed by Dr. Pettingill, noted that the cause was "acute hemorrhage of lungs" with "miner's consumption" as a contributing factor. Most likely, this is the same 60-year-old man, identified by the *Silver State* as "Charles Shreink," who "dropped dead" on the same day. A miner, Schein had been one of the first to arrive in the Gold Circle area.

George B. Clawson, 43 years old, who died on October 29, 1919. According to the *Daily Free Press*, "the deceased has been an invalid practically all his life and his death was not unexpected as he had been failing for some time." Dr. Pettingill certified influenza as the cause of death. His funeral was held in Midas on the afternoon of October 31. George Clawson was Cecilia Wagner's brother.

Fay Kennedy, approximately 35 years old, who died on February 4, 1920. According to her death certificate, which identified her as a prostitute, she died from "acute alcoholism - probably wood alcohol in lilac toilet lotion." *The Humboldt Star* explained that she died "from drinking hair tonic. It is not known whether she drank the stuff with suicidal intent or by mistake in ignorance of its poisonous character. She was in terrible agony for hours before death ended her suffering." Kennedy had been in Midas for one week prior to her death. Dr. Pettingill attended, and Frank Macy served as undertaker.

John Burroughs, 52 yeas old, who died on May 16, 1921. A miner, Burroughs had been in Midas for eight months and died of "alcoholic poisoning." R.T. "Dick" Noble served as undertaker.

Nad Uerant, 60 years old, who died on May 11, 1927. His death certificate stated that the Austrian immigrant "was blasted in the Thompson tunnel." *The Humboldt Star* speculated that he had been the victim of a dynamite blast that went off late in the mine in which he was working. He had been alone at the time and was not found for many hours after his death. He left no known survivors.

Arthur Leek, 66 years old, who died on May 6, 1930. The Battle Mountain newspaper noted that he was found in bed, dead from natural causes. The farm hand was buried two days later in the cemetery with Frank Macy serving as undertaker.

Frank Hutchinson, 65 years old, who died on January 18, 1932. He might or might not be buried in the Midas Cemetery. His death certificate noted that he died of a stroke or heart attack, but did not name the place of burial. Gordon Warren recalled the burial because of the large amount of snow in the area at the time. The Elko *Daily Free Press* implied that Mr. Hutchinson was buried in Midas by noting that " considerable difficulty was experienced at the funeral due to the depth of the snow and it had to be postponed on one occasion." The Winnemucca newspaper reported that the body would be transported to Battle Mountain, but the Battle Mountain paper was silent on that subject when it reported the death of the long-time Midas resident.

Louis Reymond, 52 years old, who died on September 21, 1932, having collapsed on the trail between Midas and the Elko Prince Mine. A mining engineer and native of Switzerland, Reymond had moved to Midas from Tonopah in 1916. The year before, a miniature mine and mill designed and built by him was displayed at the Mines and Metallurgy Building of the San Francisco Panama-Pacific Exposition. Also displayed were samples of gold ore from the Elko Prince Mine, the Rex Mine, Frank Macy, Richard T. Noble, and J.A. Paisley.

In addition to being a pilot and one of the first car owners in Midas (1918), Reymond was President of the Gold Circle Buick Mining Company. Byron Wilkerson remembered that Reymond was popular with the town's children because he always had a funny story to tell. On May 27, 1920, the *Nevada State Journal* described him as "one of Gold Circle's most popular and successful miners" when it reported his wedding in Winnemucca. "The happy couple received the congratulations of their many friends, who wish them all kinds of good luck, and are now settled down in the cosy [sic] house up the canyon which Mr. Raymond [sic] has owned for some time." By 1930, however, Reymond appeared on the census detail for Midas without a spouse. Two years later, he succumbed to heart failure; Frank Macy was the coroner.

Louis Reymond displays what may have been one of the first automobiles to arrive in Midas. (Photo courtesy of Friends of Midas)

In addition to the county death certificates, other sources provide information about others buried in the Midas cemetery:

John M. Rice, 63 years old, who died on November 3, 1914, according to the tombstone still standing in the cemetery. A native of Missouri, Rice and his wife, Emma, owned their home in Midas, according to the 1910 census. Former residents remember the Rice home as being the wood frame house that still stands, just barely, between Main Street and East Street behind homes currently belonging to the Mortara, Pincolini, and Mickelson families.

Diane Warren, a four-day-old "blue baby" whose sister, Desda, recalled her birth and death as having happened in 1915 or 1916.

Baby Boy Cropley, who was born and died on June 7, 1919, according to his birth certificate.

An unknown person who was buried in the Midas Cemetery on July 7, 1924, according to Elko County inquest records.

A 14-year-old boy whose last name was "Howard." Edna Timmons remembered the youngster's death and burial, but no records have been located.

Fire and theft took their toll on the burial ground after the town's population dwindled. Evelyn Sabin remarked that the cemetery had burned before she arrived in Midas in the late 1950s, but 10 or 12 markers remained. Forty years later, however, only one marble headstone (for Bliss) and two granite headstones (for Halsell and Rice) still stood, hidden in thick sagebrush. Clawson relatives erected new, wooden markers for Cecilia and George in the late 1980s. In 1991, the cemetery area was cleared by honor camp inmates, revealing a charred wooden marker, the remains of a wire base of a flower arrangement, and several indentations in the ground indicating graves.

In November 1995, it was discovered that the cemetery was on land owned by Barrick Gold Corporation, a mining company that owns a portion of the historic Gold Circle Mining District. A year later, the company transferred the land to Elko County, which allows, through a cooperative agreement, Friends of Midas to improve and maintain the site. Since then, many companies and volunteers have worked hard to preserve and beautify the Midas Cemetery. In 1996, Newmont Mining Corporation donated barbed wire for fencing; Conn Davis surveyed and recorded the parcel. Under the supervision of the Nevada Division of Forestry, inmates from the Winnemucca Honor Camp removed and burned the brush and weeds.

In 1997, the inmates constructed the sections for the picket fence, which was designed from Edna Timmons's memory of the original fence. Dick Rowe and Dan Bennett painted and placed the first sections. Skip Powell dug many of the first post holes. Dan and Joan Bennett planted 400 daffodil bulbs. The next summer, Artistic Fence Company, at the request of Midas Joint Venture, drilled all of the holes for the remainder of the fence. Midas Joint Venture also arranged for additional clearing work, fencing, and road construction.

The amount of land transferred from Barrick to Elko County was larger than the historical cemetery, so the county, at the request of the Friends of Midas, determined that no new burials would be allowed in the historic section. Elko County provides for new interments on the excess land. So far, only two individuals rest in the new cemetery:

Ronald D. "Arnie" Hasson, 49 years old, who died December 9, 1992. A veteran of the United States Navy, Hasson had been a fixture in Midas for several years prior to his death. The fire station built on the site of the old Midas

Town Hall was named for him in 1991.

Kenneth Dale "Kenny" Higgins, 54 years old, who died on May 13, 1995, in Reno.

Although **Ted and Usa "Kitty" Hanson** are not buried in the cemetery, a marker there commemorates the lives of these long-time Midas residents. Ted died in 1981 at the age of 63 years, and Kitty died in 1999 at 68 years of age.

Taken from the rocks above Midas Canyon, this picture shows the main part of town under a summer canopy of cottonwoods. The building with the pitched roof in the immediate foreground is George DeLong's garage, the site of many an FOM event. In the background, the Midas road can be seen meandering out to Squaw Valley where it dead-ends at the county road. (Photo courtesy of Mike Baker, Friends of Midas Collction)

FRIENDS OF MIDAS

In 1995, several Midas boosters decided to organize a formal group whose purpose would be to research and preserve the town's history. Entitled "Friends of Midas," the new organization was incorporated in Nevada and recognized by the Internal Revenue Service as a nonprofit, 501(c)(3) organization.

The idea originated with Dan Bennett who has served as the President of the Board of Directors since its inception. As such, he has been the group's primary organizer, fundraiser, collector, and spokesperson. He will be the first, however, to say that he does not do all the work himself and that is, in fact, true. The preservation of the history of Midas has benefited greatly from the enthusiastic support of hundreds of volunteers. If there were not already many people with a great deal of affection for this tiny town, no amount of cajoling could have developed the organization.

The other original members of the Board of Directors were Dana Bennett, Secretary; Gene Laucirica, Treasurer; Scott Putnam; Mert Mickelson; Bob Unger; and Paula Scheuller. Scott Putnam retired in 1999. Gene Laucirica passed away on September 12, 2003, leaving the board and Midas much too early. Bev White filled his shoes briefly until the board was joined by Pam Mortara and Brian Elmore, the new Treasurer. John Doyle of Winnemucca has provided legal assistance.

For the first three years of its existence, Friends of Midas relied on donations of items, time, and cash. It earned additional funds from recycling aluminum cans (Midas has always had a lot of beer drinkers!), selling copies of *Forward with Enthusiasm* and note cards, and hosting raffles, such as the one at the Independence Day celebrations.

In 1998, Friends of Midas became involved in the annual

Friends of Midas enjoy fun events at George DeLong's place. (Photos courtesy of Friends of Midas)

Michael Elmore and Laney Ruf are decked out for the parade. (Photo courtesy of Friends of Midas)

Jim Schennum and Shadow lead kids and quads in the 2005 Fourth of July parade. (Photo courtesy of Friends of Midas)

Ducks Unlimited dinner held every August in Midas. It proved to be a tremendously successful event to raise both funds for, and awareness of, the organization. That year, the Midas dinner, chaired by Dick Rowe, was one of the most successful Ducks Unlimited events in Nevada, and Friends of Midas benefited from that success. In 2001, the organizing committee chose to go with the Mule Deer Foundation, and in 2003, switched again to Nevada Bighorns Unlimited, which has proven to be a popular organization in Midas. Friends of Midas participated in all of these events and, up until 2006, sponsored the breakfast on Sunday morning after the big dinner and auction.

Midas enthusiasts, however, continued to ask to join the organization: there was great interest in being card-carrying "Friends of Midas." In 1998, the board voted to offer memberships at a cost of $25 each year for an individual membership and $50 for a family membership. Lifetime memberships are available at $1,000 for individuals and $2,500 for corporations. Any amount received for a lifetime membership is deposited in an investment account; only the earnings on that principal may be used for operating expenses during the member's lifetime.

Memberships were first made available on October 16, 1998, the date of the Midas School Celebration; anyone who joined that day was designated a charter member. Edna Wilkerson Timmons was honored at the ceremony as Friends of Midas Charter Member No. 1. By the time the dance ended that night, the group had 73 charter members. The most recent membership list holds 120 individual names.

Additionally, as of this writing, there are 15 individual lifetime members and four family lifetime members. Individual lifetime members are Dan Bennett, Dana Bennett, Joan Bennett, Art Cimijotti, Bob Lukens, Bev Matson, Les Matson, Pam Mortara, Paula Scheuller, Tony Scheuller, Keith Thompson, Sheri Thompson, Lori Wegren, Gus Wegren, and

Ben Tedore and Julie Mazzone had a distinctive entry in the 2006 parade. (Photo courtesy of Friends of Midas)

Daffodils in the Midas Park (Photo courtesy of Friends of Midas)

From left to right, youngest to oldest, are Frank and Bell Macy's daughters: Francis Marion Macy, Lorene Midas Macy, Irene Liberty Macy, and Minerva Delle Macy. All four were born in Midas; Minerva has the distinction of being the first baby born in the new mining camp. (Photo courtesy of the Lyon Family, Friends of Midas Collection)

In 1911, this wagon, now in the Midas Park, transported victims of the last Indian massacre in Nevada, (Photo courtesy of Friends of Midas)

This fresno, now in the Midas Park dug canals east of Midas in the 1880s. (Photo courtesy of Friends of Midas)

Bev White. Family lifetime members are the Corey and Annie Casci Family, Lynn and Rick Isaeff, the Jim and Carol Johnson Family, and Jim and Pat Swisher.

In 1999, the Board named several honorary lifetime members, due to their enthusiastic support of, and invaluable assistance to, the preservation and restoration of the history of Midas. These people had all lived in Midas during its heyday, and many had been born in Midas. In addition to Edna Timmons are Minerva Lyon Dunn, Edna Johnson Schuyler, John G. "Jack" Murdock, Noreen Wilkerson Murdock, Robert Pilant, Joyce Lyon Triplett, Byron Wilkerson, and Desda Warren Wood. Seven years later, in gratitude for many years of support and assistance, the Board named Newmont Mining Corporation an honorary corporate member.

Friends of Midas oversees the historic Midas Cemetery, the Midas Jail, and the Midas School property. In 1996, Elko County set aside 12 lots near Midas Creek as the Midas Park, a place for Friends of Midas to display old mining and ranching equipment. On July 4, 1998, Friends of Midas received a generous donation from Les and Bev Matson, then-owners of the Midas Saloon and Dinner House, for the mining exhibit at the park based on the design presented by Reno landscape architect John Priester. One of the items featured in that plan, the blacksmith shop, was built in the summer of 2006. It includes a hand-forged hasp from Dick Nietfeld and will house authentic equipment to make wagon wheels.

Water is provided by a spring in the center of the park and a supplemental connection that serves a drip irrigation system. In 2000, Friends of Midas purchased and planted five Colorado spruce trees. Originally open to local cow traffic, the park is now encircled by barbed wire fencing, which is essential to ensure the survival of the trees. Recently, trees have been planted to commemorate departed Midas friends. Currently, trees memorialize Betty Elges, Gene Mortara, and Clayton and Virginia Phillips. In 2006, trees remembering Gene Laucirica and Byron Wilkerson will be planted.

In the summer of 2004, the May Brothers Ranching Company loaned two nineteenth century ranching implements: a horse-powered hay baler and a wagon from the John G. Taylor ranching empire. Newmont Mining Corporation provided trucks and a crew to transport the equipment from the Hot Springs Ranch to the park where they may be seen for the first time in almost a century. Other pieces displayed there include a wagon used in the aftermath of the 1911 Massacre, a single bottom plow, a potato digger, a fresno used in the construction of the Willow Creek and Rock Creek canals, four tractors, a horse-drawn road grader (also on loan from the May Brothers), and miscellaneous pieces of cars and engines.

In addition to historic buildings, Friends of Midas owns three different, mostly vacant parcels of land. Ken and Ingela Snyder donated a parcel of four lots, on which is located a large metal storage structure, to the group. A parcel of two lots was purchased from the Keller Family because of its importance to Midas history. The 2004 Friends of Midas newsletter explained:

> Frank and Bell Macy were among the first to seek their fortune in the canyon in October, 1907. Before they had the opportunity to build the home they wanted, they lived in a tent, a dugout, and a makeshift tarpaper-covered wood shed. Eventually, the Macys had four daughters and established their family home at the northwest corner of the intersection of Main and Third Streets. For 30 years, Frank Macy was the Justice of the Peace and very active in mining. [He was the town barber, too, and early Midas correspondent to the *Humboldt Star*.] Family members moved on long before the house burned in the 1960s, but the property remained with the family and was most recently owned by Dale and Vicki Keller. Dale's mother was Francis M. Macy, the youngest of the four daughters.

Supplied with electricity, sewer, and water services, the parcel currently provides camping space for visitors.

The third parcel consists of 12 lots donated by Dan and Joan Bennett and is occupied by the Long Chance Mine, which did not produce much gold, but

At 6 months of age, Oscar Leonard Berg (born in Midas on January 5, 1918) enjoys some Midas front porch sunshine. (Photo courtesy of Len Berg, Friends of Midas Collection)

Oscar Leonard Berg enjoys some Midas front porch sunshine at the school in 2005. (Photo courtesy of Len Berg, Friends of Midas Collection)

did produce a good story. The nearest neighbor to the Long Chance is the Elges home, which was the site of the red-light district over 70 years ago. In a recorded interview, Edna Timmons described a fire at that location and the escape route that the Long Chance had provided for "old Theodore Caswick."

> ... it [the fire] happened real early in the morning, and he hid up in that tunnel, because he didn't want to get caught. I can remember that happening and hearing them laughing and everything about finally seeing old Theodore up there, and they had to wait until nighttime, because he didn't want anybody to see him. And somebody realized that he was in there and took clothes to him and things.

In the past decade, numerous people and several companies have given hundreds of items to Friends of Midas. Although many pieces were lost in the burning of the Midas School, most of the personal items, such as pictures and letters filled with memories of early Midas, were not. The organization has acquired a deep collection of archival material related to the development of the town and the mining district that few other Nevada historical groups can equal.

One particularly interesting donation to the collection is the Berg silver ingot. Oscar Wilhelm Berg, his wife Anna, and their two-year-old daughter Astrid settled in Midas in 1915, three years after leaving Finland. Oscar was the mill foreman at the Elko Prince from 1917 until it burned in 1922. During that time, a son, Oscar Leonard (Len), had been born, and the elder Oscar had carved his initials on a bar that consisted of 1.6 pounds of Midas silver.

The family left Midas after the mill burned, and the ingot was forgotten. Over 80 years later, in October 2003, the silver bar surfaced at a Fred Holabird's Americana Auction in Reno. No one knew what the initials O.W.B. represented until Friends of Midas, with Len's assistance, quietly and successfully purchased his father's souvenir at the Reno auction. Once the ingot was safely in the possession of Friends of Midas, Len told the story to Holabird who publicized it in his widely-read newsletter. Plans were made to display the ingot prominently in the Midas School Museum. Luckily, however, it was not there when the school burned.

This historic preservation group is unique because it is actively supported by a broad range of people and companies. In the first edition of the group's newsletter, almost 10 years ago, an attempt was made to name every person and organization who had, so far, contributed to the fledgling organization. This roll call included well over 100 names. The vast majority of those listed are still involved with Friends of Midas and have been joined by many, many others. To name them all would be a daunting task. As the September 2000 newsletter said:

> A lot of generous people have made all of these things happen much faster than ever was anticipated. It would be impossible to name all of the volunteers without overlooking someone. Attorneys, accountants, surveyors, builders, mining companies, drilling companies, legislators, state and county officials and a lot of ordinary working stiffs have given generously of their time and expertise.

These are all true Friends of Midas.

The Berg Ingot consists of 1.6 pounds of Midas silver with a touch of gold. It is 3.5 inches long and slightly over 1 inch wide. (Photos courtesy of Friends of Midas)

Still a prominent feature above Midas, this unnamed mine was more speculative than productive. (Photo courtesy of Roy Powers)

BIBLIOGRAPHY

Battle Mountain Herald. Battle Mountain, Nevada. Various issues, 1907-1911.

Battle Mountain Herald and Central Nevadan. Battle Mountain, Nevada. Various issues, 1907-1911.

Battle Mountain Scout. Battle Mountain, Nevada. Various issues, 1913-1962.

Burge, Florence, and Noble H. Getchell. "From Grease Paint to Gold: The Life of Noble Hamilton Getchell." Unpublished manuscript, Florence Burge Papers, Special Collections Department, University of Nevada, Reno, 1958.

Carlson, Helen S. *Nevada Place Names: A Geographical Dictionary.* Reno: University of Nevada Press, 1974.

Carson City Daily Appeal. Carson City, Nevada. Various issues, 1907-1930.

Central Nevadan. Battle Mountain, Nevada. Various issues, 1907.

Clawson, Arthur Raymond, editor. *Freighting Pioneers: Anecdotal History of the Clawson and Thompson Families and Their Involvement in Freighting and Mining in the Tuscarora - Independence Valley Region of Northeastern Nevada.* San Diego, California, 1991; revised, 1992.

Clawson, Marion, preparer. *History of the Clawson and Thompson Families.* 1971.

Daily Free Press. Elko, Nevada. Various issues, 1911-2006.

Daily Independent. Elko, Nevada. Various issues, 1907-1915.

Dorr, J.V.N. and Dougan, L.D. "Elko Prince Mine and Mill." Colorado meeting of the American Institute of Mining and Metallurgical Engineers. September 1918.

Dove, George R. #1051. Inmate File. Division of Archives and Records, Carson City, Nevada.

Dreisner, Doug and Coyner, Alan, The Nevada Division of Minerals. *Major Mines of Nevada: Mineral Industries in Nevada's Economy.* Reno: Nevada Bureau of Mines and Geology, annually, 1998-2006.

Elko Free Press. Elko, Nevada. Various issues, 1909-1911.

Elko Independent. Elko, Nevada. Various issues, 1915-1999.

Emmons, William H. "A Reconnaissance of Some Mining Camps in Elko, Lander, and Eureka Counties, Nevada." *U.S. Geological Survey Bulletin 408.* Washington, D.C.: Government Printing Office, 1910.

Ferguson, Henry G. "The Mining Districts of Nevada." *Economic Geology* 24. March–April 1929.

Florin, Lambert. *Nevada Ghost Towns.* Seattle: Superior Publishing Company, 1971.

Folkes, John Gregg. *Nevada's Newspapers: A Bibliography, A Compilation of Nevada History, 1854–1964.* Carson City: State Printing Office, 1964.

Ford, Victoria, ed. *Midas, Nevada: Mining, Milling, and Memories.* Reno: University of Nevada Oral History Program, 2001.

Franco-Nevada Mining Corporation, Ltd. *New Engines of Growth: 1999 Annual Report.* Toronto, 2000.

Free Press. Elko, Nevada. Various issues, 1907–1908.

Frickstad, Walter N.; Edward W. Thrall; and Ernest G. Meyers. *A Century of Nevada Post Offices, 1852–1957.* Oakland, California: Philatelic Research Society, 1958.

Friends of Midas, Inc. Collected materials about the Gold Circle Mining District and Midas. Midas, Nevada.

______. Newsletters. Midas: annually, 1996-2006.

Gallagher, Mervin, J., compiler. *Nevada Mines, Mills, and Smelters in Operation as of July 1, 1959.* Carson City: State Printing Office, 1959.

Getchell, Noble. Papers. Special Collections Department, University of Nevada, Reno.

Gold Circle Nevada: Guide and Manual. Chamber of Commerce. Midas, Elko County, Nevada. Undated.

Gold Guidebook for Nevada and Utah. Minobras Mining Services, 1981.

Granger, Arthur E., Mendell M. Bell; George C. Simmons; and Florence Lee. *Geology and Mineral Resources of Elko County, Nevada.* Reno: University of Nevada Bulletin 54, Geology and Mining Series, 1957.

Heller, Dean, Secretary of State of Nevada, Issuer. *Political History of Nevada 1996.* Tenth ed. Carson City: State Printing Office, 1996.

Hendershot, Carol. "Jess Harris: The Flying Sheriff." *Northeastern Nevada Historical Society Quarterly.* 92-2. Elko, Nevada, 1992.

Humboldt Bulletin. Winnemucca, Nevada. Various issues, 1962-1964.

Humboldt Star. Winnemucca, Nevada. Various issues,

Richard and Mary Pincolini lead a troop of festive four-wheelers.
(Photo courtesy of Friends of Midas)

1907-1925, 1935-1961.
Humboldt Star and Battle Mountain Scout. Battle Mountain and Winnemucca, Nevada. Various issues, 1963-1965.
Humboldt Sun. Winnemucca, Nevada. Various issues, 1972-2006.
Hunt, S. Frank. *Mining Geology Outlined.* 1936.
Inventory of the County Archives of Nevada. No. 4. Elko County. Reno: The Historical Records Survey, December 1938.
LaPointe, Daphne D.; Joseph V. Tingley; and Richard B. Jones. *Mineral Resources of Elko County, Nevada.* Reno: Nevada Bureau of Mines and Geology Bulletin 106, 1991.
Las Vegas Review-Journal. Las Vegas, Nevada. Various issues, 1949-2006.
Leigh, Rufus Wood. *Nevada Place Names: Their Origin and Significance.* Salt Lake City: Deseret News Press, 1964.
Lincoln, Francis Church. *Mining Districts and Mineral Resources of Nevada.* Reno: Nevada Newsletter Publishing Co., 1923. Reprinted by Nevada Publications, Las Vegas.
Lingenfelter, Richard E. and Karen Rix Gash. *The Newspapers of Nevada: A History and Bibliography, 1854–1979.* Reno: University of Nevada Press, 1984.
Macy, Edith Bell. *Day Book.* Midas, Nevada. October 8, 1907–October 18, 1908.
"Midas." *Desert.* September 1968.
"Midas Had a Mistress." *Desert.* August–September 1965.
Murbarger, Nell. *Ghosts of the Glory Trail.* Palm Desert, California: Desert Magazine Press, 1956.
Nevada Appeal. Carson City, Nevada. Various issues, 1984-2006.
Nevada Legislature. *Journal of the Assembly.* Carson City, Nevada: Legislative Counsel Bureau, biennially, 1909–2006.
______. *Journal of the Senate.* Carson City, Nevada: Legislative Counsel Bureau, biennially, 1909–2006.
______. *Statutes of Nevada.* Carson City, Nevada: Legislative Counsel Bureau, biennially, 1909–2006.
Nevada Mines Sales Opportunities 1947. Los Angeles: Board of Supervisors of LA County, 1947.
Nevada's Metal and Mineral Production (1859–1940, inclusive). University of Nevada Bulletin Geology and Mining Series 38, November 1, 1943.
Nevada. Office of State Inspector of Mines. *Biennial Report.* Carson City, Nevada: State Inspector of Mines, 1910–1972.
Nevada State Department of Education. *Educational Directory.* Carson City, Nevada: Department of Education, annually, 1908–1955.
______. *Report of the Superintendent of Public Instruction.* Carson City, Nevada: Department of Education, annually, 1908–1955.
Nevada State Journal. Reno, Nevada. Various issues, 1907-1983.
Nevada State Writers' Project. *Origin of Place Names, Nevada.* Reno: Nevada State Writers' Project, 1941.
Newmont Mining Corporation. *2004 Annual Report.* Denver: www.newmont.com, 2006.
Noel Devine vs. East Standard Mining Company. Civil Case #3470. Humboldt County Courthouse, Winnemucca, Nevada.
Official Results of General Election. Biennially, 1908–1952.
Outline of Nevada Mining History. Reno: Nevada Bureau of Mines, University of Nevada, 1964.
Paher, Stanley W. *Nevada: An Annotated Bibliography.* Las Vegas: Nevada Publishers, 1980.
_________. *Nevada Ghost Towns and Mining Camps.* Berkeley: Howell–North Books, 1970.
Patterson, Edna B.; Louise Ulph; and Victor Goodwin. *Nevada's Northeast Frontier.* Sparks, Nevada: Western Printing and Publishing Co., 1969.
_________. *Sagebrush Doctors.* Springville, Utah: Art City Publishing Co., 1972.
_________. *Who Named It? History of Elko County Place Names.* Elko, Nevada: Warren L. and Mary J. Monroe, 1964.
Payne, Anthony. *Nevada Mineral Exploration and Mine Development, 1950–1972.* Washington, D.C.: Council on Environmental Quality, 1976.
Pointing the Way: Elko County Public Schools. Nashville: George Peabody College for Teachers, 1956.
Raymond, C. Elizabeth. *George Wingfield: Owner and Operator of Nevada.* Reno: University of Nevada Press, 1992.

Reed, Waller H. *Population of Nevada Counties and Communities 1860–1980.* Reno: Nevada Historical Society, 1984.

_________. *Election Returns by County.* Reno: Nevada Historical Society, 1984.

Reno Evening Gazette. Reno, Nevada. Various issues, 1907-1983.

Reno Gazette-Journal. Reno, Nevada. Various issues, 1983-2006.

Richie, R.L., Editor. *The Directory of Nevada Mines.* Reno: The Mining Press, 1940.

Rott, Edward, Jr. *Ore Deposits of the Gold Circle Mining District, Elko County, Nevada.* Reno: University of Nevada Press, August 1931.

Salter, Chuck. "The Town That The Internet (Almost) Forgot." *Fast Company.* April 2001.

Schrader, Frank C.; Ralph W. Stone; and Samuel Sanford. "Useful Minerals of the United States." *U.S. Geological Survey Bulletin 624* (Washington, D.C.: Government Printing Office, 1917): 191, 194.

Stevens, Sue Ann. "Midas: A 20th Century Gold Mining Camp." *Northeastern Nevada Historical Society Quarterly.* Elko, Nevada. Fall 1984.

Stuart, E.E., Inspector of Mines for the State of Nevada, compiler. *Nevada's Mineral Resources.* Carson City: State Printing Office, 1909.

The Chukar Partridge in Nevada. Nevada Fish and Game Biological Bulletin #1. April 1954.

The Elko Enterprise. Elko, Nevada. Various issues, 1916-1917.

The Gold Circle Miner. Midas, Nevada. April 11, 1908.

The Gold Circle News. Midas, Nevada. June 20 and September 26, 1908.

The Gold Circle Porcupine. Midas, Nevada. May 20, 1914.

The Mine Search Annual 1983. Vol. VIII: Nevada. Boulder, Colorado: Metals Economics Group, 1984.

The Nevadan. Reno, Nevada. Various issues, 1966.

The Silver State News. Winnemucca, Nevada. Various issues, 1907-1909.

Toll, David W. *The Compleat Nevada Traveler.* Reno: University of Nevada Press, 1976.

Torrence, C.W. *History of Masonry in Nevada.* Reno: Grand

Lodge F&AM of Nevada, 1944.
Tuscarora Mining News. Tuscarora, Nevada. Various issues, 1907-1908.
U.S. Bureau of the Census, Manuscript censuses, 1910, 1920, 1930. Online. Ancestry.com.
U.S. Bureau of Mines. *Mineral Resources of the United States.* Washington, D.C.: Government Printing Office, annually, 1907–1931.
_________. *Minerals Yearbook.* Washington, D.C.: Government Printing Office, annually, 1932–1967.
_________. *Statistical Appendix to the Minerals Yearbook.* Washington, D.C.: Government Printing Office, annually, 1932–1935.
U.S. Post Office Department. *Reports of Site Locations, 1837–1950.* Microfilm, National Archives, Washington, D.C.
______. *Record of Appointment of Postmasters, 1832–September 30, 1971.* Microfilm, National Archives, Washington, D.C.
Vannette (sic), Dick. #3181. Inmate File. Division of Archives, Carson City, Nevada.
Weekly Independent. Elko, Nevada. Various issues, 1907-1914.
Wilcox, F. Wayne. *The Life and Times of Freeman Wayne Wilcox.* Undated.
Zanjani, Sally. *A Mine of Her Own: Women Prospectors in the American West, 1850-1950.* Lincoln: University of Nebraska Press, 1997.

Drawn by Lonny Davis of Elko in 1971, this portrait of the old Warren Mercantile has hung in the Midas Saloon and Dinner House for many years. (Courtesy of George and Terri Watford)

INDEX

Page numbers in italics indicate a photograph, document, or drawing of the subject in this index

About the Author

Dana Bennett would say that she splits her time between Midas and Paris, two of her favorite places in the world, but she is not a writer of fiction.

She is, however, a writer of nonfiction and has published articles about polygamy, cowgirls, legislators, and rodeo. Currently a doctoral student in public history at Arizona State University, Dana remains a Midas land baroness and proud member of the Coyote Sisters. She is distantly related to Henry "Ol' Pancake" Comstock (of Nevada's Comstock Lode fame) but more closely related to her delightful nieces, Kalli and Kaycee, and terrific nephews, Jordan and Eric, all of whom would rather be in Midas than in school. (Well, that's not quite accurate, but she is trying to set a good example for them.)

After crazy and successful careers as both a fact-finding, report-churning researcher and a whiskey-swilling, cigar-puffing lobbyist, she recently got a wild hare, quit her job, and moved to Phoenix to pursue her PhD. Certainly, there are moments when Dana would rather be in Midas than a class on historical theory (Is it postmodernism? New historicism? Probably omphaloskepsis!), but for the time being, she is studying hard at one of the most forward-looking and enthusiastic universities in the country.

It is true: her all-time favorite Midas beverage is rum and Diet Coke®, but a well-stirred Manhattan runs a close second. Luckily, both erase brain cells; otherwise, Dana would have even more interesting tales to tell about life in Midas, Nevada.

About the Designer

Nancy Peppin is a book designer and watercolor painter, known for her images of Twinkies in odd settings. She also paints still-lifes (of the contents of drawers, closets, refrigerators and other containers) and landscapes of Nevada. Peppin has been a painting and drawing instructor in the art departments at Truckee Meadow Community College and the University of Nevada-Reno, and she is a graphic designer in the slot machine industry.

Her career has included museum and exhibit design and graphic art for the Great Basin Chautauqua. In 1992, she painted a series of five murals portraying the fictitious back rooms of Nevada museums and public buildings for the Downtown Reno Redevelopment Project which were attached to the Riverside Hotel for five years. In 2001 she painted the poster for the Great Reno Balloon Race. She has exhibited for more than 25 years in the northern Nevada area and has paintings and commissions in many collections.

Peppin has lived in Reno since 1974, coming from the San Francisco Bay area. She has been active in several community organizations, in particular, public television, where she has worn funny hats on live TV to induce people to donate. Companion cats and collections of moose, books, and orchid plants have taken over her house and hold her hostage.